WATER
FOR EVERY FARM

USING THE
KEYLINE PLAN

WATER
FOR EVERY FARM

USING THE
KEYLINE PLAN

By P.A. YEOMANS

Published in 1981 by Second Back Row Press Pty Limited,
50 Govett Street, Katoomba, 2780.

'Water For Every Farm' first published by
Murray Books 1965.
Reprinted 1968.

'The Keyline Plan' first published by P.A.
Yeomans 1954.

'Water for Every Farm' and 'The Keyline Plan'
reprinted as 'Water for Every Farm'
by Murray Books, 1978.

Copyright © P.A. Yeomans 1978, 1981

Printed at Griffin Press Limited, Adelaide.

ISBN 0 90932 529 4

PUBLISHER'S NOTE

The two books incorporated within these covers are probably two of the most important to be published in Australia.

To re-issue them at a time when Australia's rural industry is at its lowest ebb for years, with little sign of a resurgence, might seem odd, but the publishers believe that now — when much of the country has just passed through a massive drought, after several bad years with little rain — is the time to bring up again the once-revolutionary principles embodied in this book.

When P.A. Yeomans first propounded them, and demonstrated not only their practicability but also their efficiency, the ideas were regarded as revolutionary, but since then they have been widely adopted in many of the drier parts of the world, particularly in the US, Africa, the Middle East, and even in the Soviet Union. Only in the land of their birth have the ideas been either ignored or adopted by only a small number of landholders.

The dry continent has too long been accepted as something which has to be put up with — about which nothing can be done.

That that attitude is wrong is easily shown, but whether the slowly grinding mills of officialdom will ever implement the ideas herein only time will tell. But time is not on our side, and it may well be left to the individual farmers, with their futures at stake, to take up the challenge and adopt methods of water conservation which the rest of the world in similar circumstances now takes for granted as a normal part of farming technique to counter the effects of drought.

First part of the book is basically the theory of Keyline, from the book The Keyline Plan, with a brief outline of its application.

Second part is a look in greater depth at the theory, and then the application of its points, spelled out so every farmer can do his own planning and work, without expensive consultants.

The whole work is one which is long overdue for the most widespread recognition on official level, with concomitant changes in official thinking on both soil development and water conservation, with suitable tax rebates to provide incentives for the work to be undertaken.

The Publisher

Contents

PART ONE

The Keyline Plan is copyrighted and some of the items mentioned may be protected by patents and design registrations.

However, any farmer has permission to use for his own purposes on his own land any of the ideas and techniques of this or other Keyline books.

P. A. Yeomans.

CHAPTER ONE

The philosophy of Keyline

KEYLINE IS A SET OF PRINCIPLES, techniques and systems co-ordinated into a plan for the development of farm and grazing landscapes — a master plan for the elaboration of a "replacement" for the natural or existing landscape.

A principal aim of Keyline is to increase both the depth and the fertility of the soil so that the soil of farming and grazing land is safe and permanent and capable of continuous improvement. Thus productiveness of the soil is increased and the quality of its products is improved.

Keyline seeks to remould the landscape, first, by the proper assessment and appraisal of all the natural and renewing resources of each individual area on which it is applied and second, by special methods of planning, design based on water control, and land management, to use those resources to the utmost for bettering the soil.

Soil, the raw material of all agricultural pursuits, owes its existence and its state of fertility and productiveness to many factors. The principal circumstances which have determined a particular natural soil are (1) the minerological and structural framework, (2) the prevailing climate, and (3) the soil's biotic associations. Likewise those three factors in combination determined the type of the natural landscape which contained or housed the original soil.

All present farming and grazing practices have involved a drastic interference with the previous landscape. It is the manner of that interference which is of the utmost concern in Keyline.

Although the natural order for any tract of land is one of continuous change, there was, in most natural landscapes which have later had any agricultural significance, a certain stability which tended to preserve the general shape and form of the land against rapid change and remoulding of the surface. But in Western man's interference with the differing natural regions to convert them into farming and grazing land that stability has too often been lost. The result has been a rapid decline in the fertility of the soil, much of which was poor to begin with, followed by soil erosion in its various destructive and well publicised forms.

However, by improved methods of planning and design based on the proper assessment of the resources of the land, man can interfere with the various natural landscapes in a beneficial way. At the same time and quite automatically he can exclude any possibility of soil erosion. The soil can be improved to a state of fertility well above that which had ever previously existed and the replacement panorama can be more stable and in better balance than was the preceding one. The quantity and quality of production can be greatly superior to that from present methods of land conversion, including what has of late been given the rather futile title of conservation farming.

Of the three factors which have determined the soil type (1) the minerological and structural framework is predetermined and not readily altered, (2) the climatic background is likewise fixed and not capable of control by man, and (3) the biotic association is susceptible to control.

While the climatic background (which may be arctic, temperate or tropical, maritime or continental and may be affected by large bodies of water and rivers, by hills, mountains and valleys and other local variations) is fixed, the particular effect which the prevailing general climate has on the climate within the soil is accessible to great changes. Thus man is given the chance to remake the soil.

Soil has a climate of its own which is absolutely crucial to the soil's teeming communities. Changes in the climate within the soil are therefore equally as critical for the improvement of the soils or for its impoverishment even to the point of destruction and removal.

Likewise, within the general climatic pattern, the concordantly planned and nurtured landscape of a farm may be given a climate of its own which is more favorable than that of the surrounding area. For instance, the climate near lakes can be decidedly better than the climate of the wider region, as it may be on a farm which has had its water resources developed for storage in farm dams. The irrigation of dry areas with water brought from a distant supply is also a notable example of manmade soil and landscape betterment.

Keyline plans the evolution of the replacement landscape on the two most permanent features of the natural landscape, (1) the existing shape and form of the land, and (2) the climate, which in large measure has moulded and determined its present topography.

Of the various elements of climate — heat and cold, wind, and water from local rainfall as well as from lakes and streams — the factor of water is the one most susceptible to control. Therefore to assess the water resources of any area, farm waters are divided into four categories so that, for Keyline purposes, the water resources may be better understood and evaluated.

In circumstances where water from rainfall is generally scarce, while at other times there is runoff from rainfall, the plan for the control of water involves earth works such as channels and drains, and for walls of storage dams. Such new water lines become permanent features of the new landscape and have precedence in design over other and less critical components of the planned landscape.

Because of the physical properties and behavior of water, those and all other waterlines are critical and invariably become the framework to the Keyline plan for the land. An that still applies even where water is in excess supply. All other constructional aspects of planning, such as roads, treeplanting locations for farm buildings and working paddocks, and subdivisions for crops, pastures and livestock, are fitted into the basic waterlines framework.

But as the course of water is determined by the land form over which it flows, all the waterlines of land are individual to each property because of the endless variations in their topography. Therefore, to fully understand and assess and plan water control and water movements, the various component shapes of land have been classified in the Geography of Keyline.

Again, within each of the geographical classes there is a typical pattern in the contours of the shapes of the land. The Geometry of Keyline permits the use of those patterns by disclosing techniques based on them for the better use in the soil of direct rainfall and for the even-spreading of irrigation water. Each of those functions also form part of the soil improvement techniques of Keyline.

The Keyline classification of farm waters, the geography and geometry of Keyline, together with the planning and design principles and the soil building and the constructional techniques which are based on them, are all fundamentals of the Keyline plan.

While water is the principal aspect of climate where improvements of the soil's climate and of the confined landscape's climate may be effected, there are other features of the general climatic pattern which may be modified in the re-formed or revised landscape. Notably also the damaging and moisture-robbing power of drying winds may be appreciably reduced by the planting of tree belts in locations chosen for that purpose.

The third factor of soil determination, the biotic association, is almost completely under the control of man in the unfolding landscape. While it is true that the first result from Keyline techniques for enriching the soil will be seen in the more luxuriant growth of plant life, many plants themselves have the faculty of improving the soil and further modifying the soil-climate for continued soil improvement. It is fortunate indeed that so many of those plants are of very high nutritive value in themselves. They include in their benevolent and beneficent array many species suitable for pastures, especially the legumes and other deep rooting grasses and herbs. Likewise many vegetables, which by themselves or in close association with other specific plants, have amazing soil building properties. There is also a wide variety of tree and shrub species which will assist with the job.

In widely differing climatic circumstances, the planting and the proper management of particular plant species are alone eminently practical in producing dramatic improvement in the fertility and productiveness of even the most unpromising earths. Subsoils, remaining on farm and grazing properties after the real soil has been lost by erosion, have often been transformed into fertile and highly productive soil in two or three years. The conversion can be accomplished economically and simply by Keyline cultivation methods with the sowing and later, management of selected pasture species.

If soil be the basic agricultural raw material then the properly planned enclosed landscape must be the permanent home for any really worthwhile soil.

But conversion is not enough; if soil be the basic raw material of agriculture then the properly planned and developed enclosed landscape, must necessarily be the permanent home for any really worthwhile soil.

The basic idea which I have attempted to convey throughout my work is continuity; that for all time the soil, like the crops and the livestock which it produces, persists and develops as thinking man assists nature by using and accelerating her own timeless processes. He will do it for his own full life enrichment but later generations must also benefit since, after all, agriculture is the human betterment of a continuous natural process. Of all the men who work on the land the grazier lives closest to those and has the widest opportunities for improving the soil and imprinting his own image on the whole landscape. Is not the grass the great healer of land and do not his stock provide the accelerators for the soil-forming process?

That philosophy has not been the characteristic feature of African, Canadian, American or Australian philosophy, much to the detriment of these countries. The attitude of those nations has changed, and a need for a permanent agricultural tradition is now being fostered.

CHAPTER TWO

Keyline — a new principle

THE KEYLINE PLAN embraces a system of progressive fertile soil development for all crop and pasture lands as well as for the steeper and rougher lands that have never before been capable of fast, economic improvement.

Its primary aim is the development of better soil structure, increased soil fertility and greater actual depth of fertile soil. It includes new cultivation techniques; a method of farm subdivision and layout; planning for timber and scrub clearing and water conservation and irrigation. All are planned to facilitate or assist in the production of fertile soil.

The Keyline plan is based primarily on a particular line or lines called Keyline. The lines and others related to them are used in all land development planning and act as guides for farm working.

The first aim of Keyline is to provide simple means of conserving all the rain that falls on the land into the soil itself, retard its evaporation rate and use the conserved moisture for the rapid production of soil fertility over both small and large areas of land.

The simplest form of a Keyline is illustrated in Map 1, which shows a valley formation by means of contour lines. The 54m contour line is the Keyline of this simple valley area.

The Keyline conception itself is a little technical, and an explanation of what the basic idea involves is given first.

A Keyline is a level or sloping line extended in both directions from a certain point in a valley, called the Keypoint. It marks or divides the two types of relationship, always in the same vertical interval, that a valley bears to its adjacent ridges. In one of the relationships, that above the Keyline, the valley will be narrower and steeper generally than the adjacent ridges on either side of it. In the second relationship, existing below the Keyline, the valley will be wider and

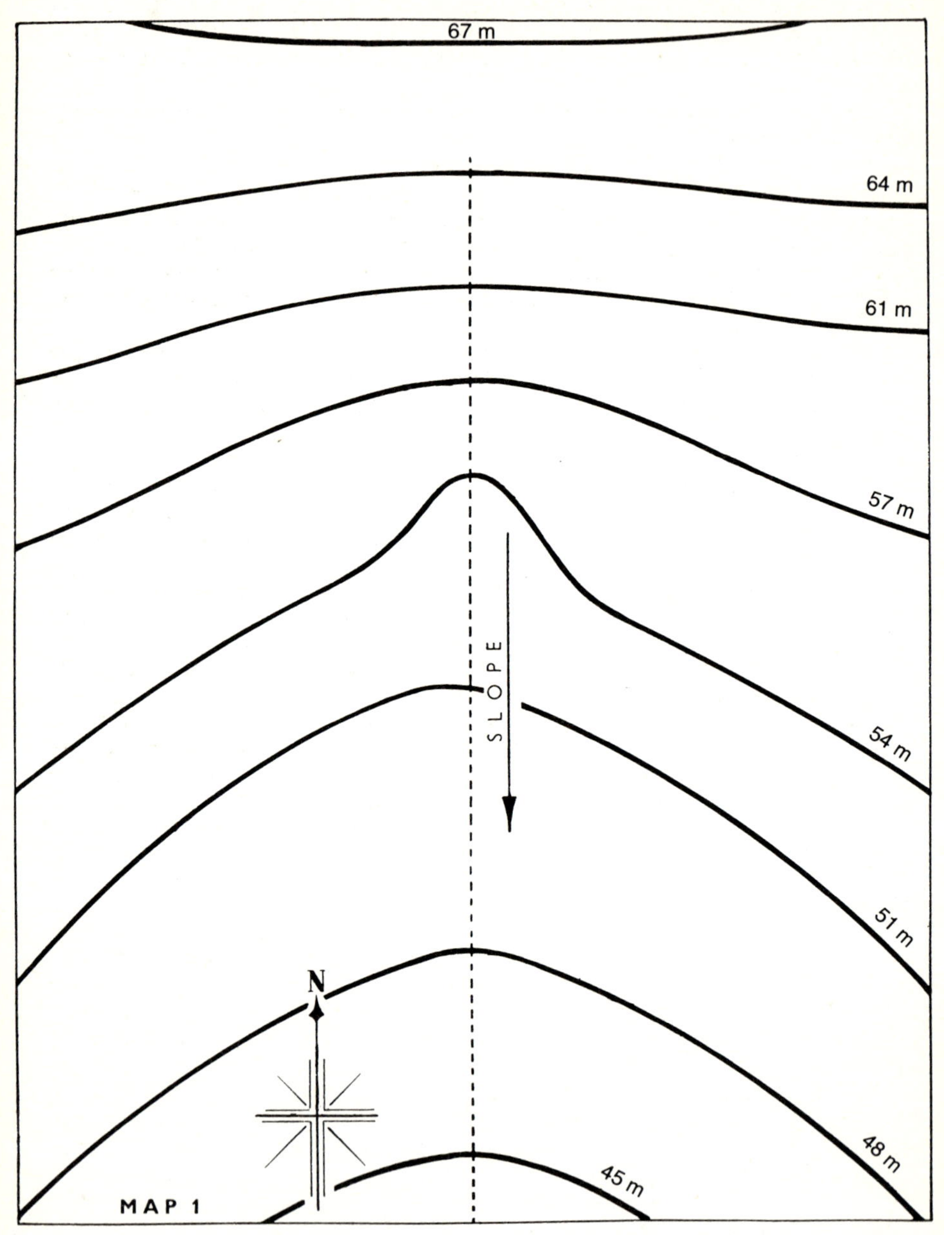

67 m
64 m
61 m
57 m
54 m
51 m
48 m
45 m
SLOPE
N
MAP 1

flatter than its immediately adjacent ridges or shoulders.

The approximate point of that relationship change in the valley is the Keypoint of the valley. A line, either a true contour in both directions from that point, or a gently sloping line rising in one direction and falling in the other direction (see later chapter) — from the Keypoint is the Keyline of that valley area.

Any property which includes in its area a watershed or water divide has one or more Keylines.

To understand the full development and uses of this and other Keylines, reference will be made to contour maps and particular contour lines of the maps. Not all readers will have had experience of these maps and their contour lines, but the following description will make the later references clear.

Contour lines, or contours, are lines on maps or marked on the land itself to show particular levels. Map 1 is a simple contour map and the contour lines on the map mark the levels.

All points on the lines marked with the various heights are the same height as indicated by the figures. Thus on the 61m contour line all points are 61m above "datum". Datum is very often mean sea-level, but may be any other permanent point.

A contour line lies at right angles to the slope of the land; as the slope changes direction the contour lines curve and turn. Contour lines on a contour map are placed at regular vertical heights apart. The distance apart is called the vertical interval. On farm contour maps these range from 7.6m to 0.6m, according to the type of land formation and accuracy desired. On Map 1 they are 3m apart vertically. The space or interval between two contour lines is referred to as a contour strip. A contour map exhibits the formation of land by means of contour lines.

The contour Map 1 exhibits a simple valley formation. The centre line of the valley floor is indicated by a dotted line and the downhill slope by an arrow.

The 6.7m contour is near the top of a watershed or water divide. The valley formation starts between the 64 and 61m contours, as indicated by the two contour lines coming closer together near the dotted line of the valley. The actual slope here is steeper than that on either side between the same two contour lines. That is the valley head. The valley steepens a little more between the 61 and 57m contours, as indicated by the two lines being closer together than the 64m and 61m contours. The slope of the valley then remains constant to the 54m contour in the valley. It is indicated on the map by the distances between the 61 and 57m contours and between the 57 and 54m contour lines at the centre valley point being about equal. At that point, where the 54m contour line crosses the dotted line of the valley bottom, a change takes place in the character of the valley formation. The valley bottom flattens considerably, as indicated by the greatly increased distance in the valley bottom between contour lines 54 and 51m.

The whole relationship of the valley to its adjacent ridges in each contour strip has also changed. Above the 54m contour line the valley bottom is steeper and narrower than its adjacent ridges in the contour strips, but below that contour line the valley is flatter and wider, in the contour strips, than its adjacent ridges. The slope relationship between the valley and the adjacent ridges continues through the lower contour strips of the map.

As a general rule, the relationship is constant for the remainder of a valley. The line of the change of relationship between the valley and its adjacent ridges in each contour strip is the Keyline of that valley. The position or point of the change in the valley itself is the Keypoint of the valley.

My own discovery, study and use of the peculiar significance relating to the varying valley and ridge forms, is the basis of the Keyline plan. Its use in farming and general land planning and development is discussed throughout both parts of this book. A study of the topographical geography of general land formation will show a remarkable consistency and regularity in the changing relationship between valleys and their adjacent ridges.

The crucial point of change in the valley floor slope, the Keypoint, may coincide with the confluence of two or more valleys.

At the Keyline the line of the valley floor and adjacent ridge slope are neutral. Various types of land formations lend differing forms to their Keylines, but generally the significant valley and ridge relationship is consistent in the widest variations of land formations.

It is important to keep in mind that the valley area in the contour strip above the Keyline is narrower generally than the adjacent ridge area and that the valley area in any contour strip below the Keyline is wider generally than the adjacent ridge area of the same contour strip.

The 54m contour line of Map 1 is the simplest form of Keyline — the Keyline of a single valley. Keylines, as discussed here for farm work, are not located on the very small scale contour maps of large land areas, such as inch-to-the-mile land plans. Maps that have sufficient contours to exhibit accurately every valley on a medium size property will, however, enable the Keylines to be located quite clearly.

Before explaining the full development of Keyline, this simple form is used in the next chapter to illustrate a practical application of the Keyline principle.

CHAPTER THREE

Absorption — the first need

THE USE OF KEYLINE as a guide or design for cultivation is discussed in this chapter. Keyline is a complete planning guide for farm development. It would seem that an overall picture of the plan should come before the details of Keyline techniques. It would, however, involve so much discussion and digression to explain new terms that I propose to present the various factors which make up the complete plan in the order that appears best for the sake of clarity. The order may not be in the proper sequence of events as they would be applied in practice.

As the various methods which make up the complete Keyline plan affect and react on each other, some repetition is necessary.

Keyline cultivation is simply cultivating parallel to the Keyline

In the various methods of cultivation of the soil to prepare land for sowing, several "workings" may be used. A "working" is a complete covering of the land area at one time with whatever implement is in use.

If one cultivation only is to be done, the single working parallels the Keyline moving away from it. Cultivation that requires more than one working to complete it is done parallel to the Keyline on the last working only.

Map 2 is identical to Map 1 except that the Keyline, the 54m contour, has been emphasised and parallel lines have been added. These lines illustrate the parallel furrows of Keyline cultivation. The parallel lines of Map 2 are drawn parallelling the Keyline.

Above the Keyline these lines parallel the Keyline moving away from the Keyline up the slope of the land. Below the Keyline they parallel the Keyline moving down the slope of the land.

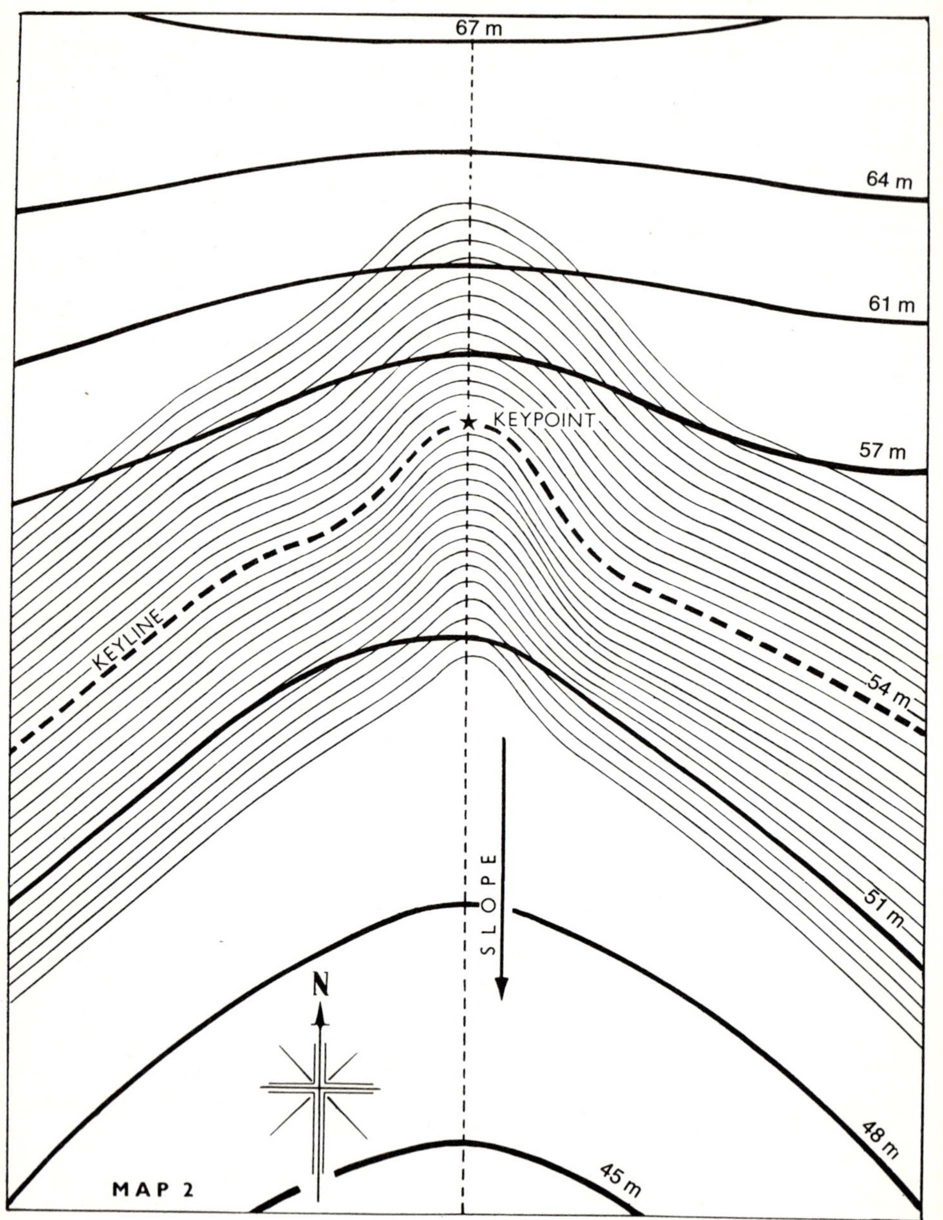

67 m
64 m
61 m
57 m
KEYPOINT
54 m
KEYLINE
51 m
SLOPE
N
48 m
45 m
MAP 2

Study of the parallel lines shows that above the Keyline they do not evenly "cut-out" the valley and ridge slopes in the first contour strip 57m-54m (190ft-180ft). The valley section is "cut-out" before the ridge sections on either side of it.

Those lines represent parallel Keyline cultivation runs, working away from the Keyline up the slope. When the cultivation lines reach the 57m contour in the valley they are some distance from and below the same contour lines on the two adjacent ridges. They have reached a greater vertical height in the valley than on the adjacent ridges. The parallel cultivation lines which started on the level or contour at the Keyline are higher in the valley than on the adjacent ridges. They slope downwards from the steeper sloping valley to the flatter sloping ridges on each side.

The parallel cultivation is continued to the upward limit of the area or paddock. When the cultivation reaches that point, there will be parts left unworked. They are cultivated out in any convenient manner without reference to the Keyline or parallel working. Their influence will not alter the effectiveness of Keyline cultivation.

Below the Keyline the parallel lines of Map 2 start at the Keyline and parallel below the Keyline down the slope of the land. The cultivation they represent does not evenly "cut out" the valley and ridge slopes in the first contour strip, that is from the Keyline to the 51m contour line. They reach the 51m contour on the ridge slopes first while the run in the valley is some distance from and above the 54m contour line in the valley. The cultivation runs are again generally higher in the valley than on the adjacent ridges in the same contour strip. They also have the same downward slope out of the valley to the adjacent ridges, as the cultivation above the Keyline. The slope of the cultivation furrows is now from the flatter sloping valley to the steeper sloping ridges, whereas above the Keyline the slope is from the steeper sloping valley to the flatter sloping ridges.

The cultivation of the area below the Keyline is completed by continuing the parallel cultivation downward to the boundary. When that is reached, areas not completely cut out are cultivated in any convenient manner. Again their influence will not alter the effectiveness of Keyline cultivation.

The significance of Keyline cultivation is apparent when two factors are considered:

1. Rainfall on or near a valley rapidly concentrates in the valley and flows off the area, not only preventing the ridges from absorbing their fair share of the rainfall, but in poor soil, taking with it some of the soil from both valley and ridge.
2. Keyline cultivation is in effect many hundreds or thousands of very small absorbent drains, preventing rainfall from concentrating in the valley — thus resisting and offsetting the natural rapid concentration of this water into the valleys.

Very heavy rainfall, after it has completely saturated the soil which has been cultivated in that way, naturally starts to move to its normal concentration lines in the valley. But it is interrupted by the tendency of almost every cultivation furrow to impede it and drift it away from the valley. The flow movement of excess water is widened and its flow is kept very shallow. The necessary time of concentration is increased enormously, thus holding the water on the land longer. The land will have time to absorb the rain that falls on it. Rainfall of maximum intensity is robbed of its destructive violence.

Keyline cultivation is completed in the order-already discussed. Cultivation above the Keyline is first completed to enable land, usually the steeper areas, to absorb the maximum or all the rain that falls on it. That prevents rapid and concentrated run off on to the flatter slope country and so protects all the land from water damage. The general result is even absorption of rainfall over the whole surface of steep land, similar in effect to the absorption of rainfall on flat, fertile, absorbent land.

This part of the significance of Keyline as a cultivation guide has been illustrated on the map with reference to contour lines both above and below the Keyline. The only need for these contour lincs is for the sake of simpler presentation. *The Keyline is the only line which is necessarily marked on the land area represented by Maps 1 and 2 for the practical application of Keyline cultivation.*

On large areas of long slope country where, for some reason, continuance indefinitely downward of the Keyline parallel cultivation is undesirable, a line is used to terminate one cultivation area and form the boundary for another.

That line is called a Guideline and is usually a true contour line, marked at a suitable distance below the Keyline. It may be a quarter mile or much farther below. The area below the Guideline is Keyline cultivated from the Guideline parallelling it downward.

Any contour line below the Keyline can be used as a cultivation guide by simply Keyline cultivating from the line downward. The effect of Keyline's cultivation diffusing and even spreading of rainfall is still completely effective.

For Keyline cultivation a special implement is needed, which properly follows the new working lines and for other significant reasons. They are discussed later.

Although the Keyline as illustrated in Map 2 is a contour line, Keyline cultivation is not strictly contour cultivation. It is rather an "off the contour" type of cultivation, which in no small measure depends for its effectiveness on the planned drift away from the valleys.

Keyline diffuses rainfall evenly over the whole of the land to absorb it in the greatest water conservation storage area — the land itself.

The field of application of Keyline extends in scope greatly from our simple first principle now presented.

Keyline planning can be applied on an area of virgin grassland or forest to develop it into a farming or grazing property. In timbered country it plans the clearing to retain timber in the best places; it positions the house or homestead, all other farm buildings, entrance and farm roads, large and small paddocks, dam sites and irrigation areas. It guides the whole course and sequence of development as well as the details of all cultivation for soil fertility improvement and high yields.

It can be applied as a planning guide to the layout and development of a public part or to the further improvement of a fully developed wheat farm or a fine grazing or dairy property.

Occasionally a very large property may have two sets of Keylines, but generally such wider applications are outside the scope of farming and the first part of this book, applying only to such developmental projects as the entire watershed of a river system.

Keyline will apply to a single small or large paddock of a farm or to land partly destroyed by erosion.

Although it cures and prevents soil erosion, that is incidental to its purpose — the development of fertile soil by the factor of absorption.

CHAPTER FOUR

Fertility —
the dominant factor

BEFORE EXTENDING THE APPLICATION of the Keyline beyond the first simple Keyline of a valley and its uses as a cultivation guide, a discussion of soil and of cultivation methods is undertaken in this chapter. It forms a basis for the presentation of the Keyline methods of progressive soil development.

Before the introduction of the mouldboard plough one of the great problems of agriculture arose from farmers' difficulties in controlling the unwanted growth on fertile soil. The rich agricultural land obtained by clearing virgin forest areas or breaking up the natural fertile grasslands was hard to hold from the exuberant growth of vegetation. The growth made it impossible for the farmer to crop large areas.

The mouldboard plough, by turning over the soil and burying the unwanted growth, gave the farmer better control. He could then hold and crop larger areas of land.

The advantage of the new power cultivation which was later introduced, lay in the further increased speed to control the unwanted growth. Rubbish was turned under to produce a "clean" soil surface.

After the earlier slow work to control the growth the new implement inspired a fetish for "cleanness" and "fineness" of cultivation. The fine seed-bed, almost universally acclaimed, produced bumper crops year after year and the rich fertile earth showed little evidence of fertility losses over long years.

That type of cultivation and the other farming and grazing methods however, were generally destroying natural fertility much faster than the crops which were profitably extracting some of it.

Eventually, when erosion became a serious menace, some nations undertook an inventory of their soil losses and found that the figures were staggering.

Gigantic efforts were needed to arrest such colossal losses by erosion. Fertile soil was not being washed away, but only those soils which had already lost or were then rapidly losing their fertility were on the move.

Fertile soil was built originally by processes of absorption, growth and decay, and such soil resists erosion. A change of methods from those which extract fertility from the soil to others which absorb fertility into the soil is the only way to overcome the erosion problem. A positive change must be made from extraction fertility farming to absorption fertility farming.

The first requirement, already stated, is the retention of all rainfall in the land for the production of fertility, and not methods to "safely" allow water to leave the property.

It is economically unsound merely to prevent erosion losses of poor soil. Soil fertility can be built back into the soil in a positive manner so much faster than the natural fertility was lost, that little need be done from the negative stand-point of controlling erosion. *The best methods of soil development are the surest means of erosion control.* Continuance of those methods will quickly produce as good, if not better, soil than that which originally existed.

While those methods are being followed, even from the first year, better farm yields will result. Absorption, growth and decay make fertile soil, and the factors which produce the maximum growth and decay can be controlled in farming practices. The needs of the farmer are satisfied at the same time.

There is little evidence anywhere in nature to support the "take and put" theory of farming where farmers are taught to "put back" into the soil each year what they "take out" in crops. So much of what is taken out is composed of materials that are available in unlimited supply from the sun, air and moisture — moisture alone requiring conservation — that if farmers cease to "mine" the top millimetres of the soil and farm the land, little if anything else need be put back. Fertilisers should be used when they are necessary, but they are rarely the "first" need. This is true of most of our farming and grazing lands.

Correct cultivation is a means of progressively improving soil structure and soil fertility, thereby developing a greater depth of fertile soil. Better crop production is incidental to the process.

The mechanics of the process of soil development where nature built up the great fertile soil belts of the earth are now reasonably well understood by the farmers. Good writers have made of the process an absorbing and fascinating story. Some see in it a miraculous efficiency and give estimates of the time required to build 25mm of fertile soil — varying from a few hundred years to ten thousand.

If the natural process is efficient and the time estimates of even a few hundred years are correct, there is little that could be done by us in the production of soil. However, nature's methods do not take time into any serious account, whereas to us "time" is all important.

The processes which developed natural fertile soil are capable of control and tremendous acceleration. The dead stalks of plants, slowly laid down by nature loosely on the land surface, decay. That is one fertility process capable of acceleration. Each time decaying vegetable matter dries, decay temporarily ceases and fertility processes are slowed down. Processes of decay are increased when moisture is present. That decay, to all intents and purposes, is fertility.

Man and his machines can stimulate decay and growth tremendously. When vegetation is stirred in the aerated part of the soil, decay continues for a longer period. Moisture remains longer to supply the needs of decay.

Every process and activity in the improvement of soil can be controlled and increased by the farmer, to the continual betterment of his soil.

Not all natural soils are fertile — far from it. Where suitable moisture, heat conditions and minerals exist, fertile soil develops in time. There is a certain progression in the development of soil in nature. The growth and decay of primary and simple forms of plant life eventually create conditions suitable for the growth of better crops and grasses.

Through the lack of some essential, the process toward the development of fertile soil will cease, or slow down. Thus poor natural soils exist in many areas.

Vast areas of those poor soils of nature can be made fertile and productive, by supplying the needs to complete their full cycle of development.

Natural shortages of vital minerals often can be remedied economically. Rainfall or other moisture sources can be controlled efficiently, to promote more rapid growth and decay. Great improvement will be made in many of those soils in a year or two. New plants and grasses which will continue and complete a cycle of high fertility can be introduced.

Plants draw their sustenance mostly from the products of decay, from and with moisture contained as a water film in the "pore" space of the soil. Generally, maximum pore space promotes maximum growth by the greater availability of pore space moisture. The pore space is multiplied by increasing the supply of vegetation for decay and for the production of humus.

Those vitally important factors are increased also by the correct mechanical mixing of vegetation into the surface soil. Correct aeration of the deeper soil and subsoil will progressively convert them to deeper fertile soil.

Some soil scientists estimate that there are 175 tonnes of living organisms and other life in a hectare of fertile soil. Those organisms generally work towards man's health and well-being.

The importance of 14, 12.5-tonne truck loads of microbes in a hectare is overshadowed completely by a four or five sheep to the hectare. The sheep or cattle obviously need constant care, but surely that other "livestock" warrants some conscious thought when it is so vital. All the elements of growth are made available to us by the various processes of the life cycles of this "life in the soil". Soil management can reduce the dynamic force to a low ebb, or tremendously stimulate its activities.

Fertile topsoil and even very poor soil can be treated as a yeast. Fed and cared for, it increases. Starved and asphyxiated, it dies.

Processes of decay are the multiplication of soil life. The processes initiate or start in the presence of moisture, air and heat. All three are necessary. That suggests a starting point in soil development should be a critical examination of farming practices and their effect on those factors.

Past cultivation habits have destroyed soil fertility to the stage where vast quantities of once valuable soil have been lost by destructive erosion. Pounding and pulverising, thining and slicing implements have all interfered with and reduced pore space in fertile soil. Soil suffered too much cultivation each time it was worked.

Extremely fine "seed-beds" are still produced on some farms, almost as if the crop in its growth was expected to devour every fine soil particle. Too fine a cultivation destroys the soil's structure, smothers and reduces soil life, thus degenerating the art of soil management into a bandit-like process of fertility extraction.

Soil fertility need not be "extracted" or destroyed to produce good crops. Crop production is properly a part of an important method in the development of better soil.

Cultivation can be either the mammoth destroyer of soil fertility or the greatest single means of improving and even the creating of more fertile soil.

An understanding of the structure and condition of naturally fertile soil and an appreciation of just what is happening, or has already happened, on some major soil areas will indicate logical means of improvement.

Fertile soil is loose, absorbent and pleasant smelling, It is dark in color, caused by decay in the production of humus. It receives rain quickly and allows it to penetrate deeply. It hold moisture in pore spaces which are found in and around the mineral particles of the soil. Moisture dries out of fertile soil slowly from the effect of the highly insulating structure of its surface. Deep soil and subsoil moisture is protected from the drying effects of sun and winds.

There are no definite horizons to the top soil, deep soil and subsoil; one merges gradually into the other and all are subject to a gentle stirring action from the larger forms of soil life and from the action of deep roots which bring nutrients to the surface. There is no sharply defined plant root zone in natural fertile soil. Shallow, medium and deep root growths mingle. Root decay acts to aerate the soil to an appreciable depth *via* the cavities left by the roots after decay.

CHAPTER FIVE

Keyline absorption — fertility

KEYLINE ABSORPTION-FERTILITY CULTIVATION techniques are so different in their effects on the land cultivated from those generally employed that their introduction on farming land will be considered as a "conversion" of land to new principles. The first year in which these new principles are used is called "conversion year".

The "conversion" cultivation has as its object the maximum possible improvement in soil structure, soil fertility and increased soil depth which can be obtained from the conversion. For that reason the first application of the technique will be different in some respects from the continuous later process. The conversion stage is to be profitable, much more so than extraction fertility methods. The continuous processes of progressive soil development are profitable both from the increase in quality and quantity of production and in the capital value of the improving land.

Soil erosion is not considered as a problem in the process, simply because it is cured incidentally. There is no problem of erosion when its cure or solution is made profitable to those directly concerned. Ordinary crop land is discussed first.

The considerations of the last chapter indicate quite clearly some important details of the type of cultivation desired.

The first requirement is minimum surface cultivation, mixing whatever vegetation is available into the few top centimetres of the soil. Some subsoil or deep soil is to be broken to provide capacity for rapid moisture absorption. With oxygen and the other vital elements absorbed, some of the subsoil is converted to live soil. That deeper soil is only broken, and none of it is brought to the surface. The deeper cultivation is to leave an uneven bottom, not all cut out clean at the maximum depth of cultivation. The cultivation is to again unite the soil into a complete structure — not a topsoil divorced from the deep soil by a compacted layer.

The surface of the cultivation is rough, rather than fine, to resist the sealing effect of heavy rain and to allow the rainfall to penetrate quickly and deeply.

The finer materials of the surface cultivation lie below the level of that rough surface. Surface wind velocity is thus reduced — moisture losses by evaporation are lowered.

The deeper cultivation conforms to the Keyline cultivation, which holds excessive rain longer on the land and permits more complete absorption.

Rainfall is quickly absorbed into cultivated poor land, making it wet and heavy on the rough, uneven bottom of the cultivation. The heavy wet soil is effectively knitted to the land and resists substantial water flow if it occurs.

New Keyline absorption-fertility cultivation is practically erosion proof; within a year or two of the resulting improvement to the soil, it is certainly so. The maximum depth of the conversion year cultivation requires some serious consideration.

In so-called shallow or thin soil, the cultivation is restricted to a depth which can be converted successfully to an improved structure by the aid of the fertility in the top soil. Considering the top fertile soil as a yeast, it is likely that too deep a cultivation could restrict the rate of soil development. That happens if a large amount of vegetation is not available for stirring into the surface cultivation. That type of soil rarely has a large amount of vegetation available.

The too deep cultivation of sticky clay subsoil is a waste of time and money. It will seal immediately rain falls. There is generally little purpose and no profit in cultivating to depths which cannot be held by definite soil improvement.

A good general depth guide for conversion year cultivation is double the depth of previous ploughing for crop productions, that is, about 203mm deep, and in the poorer soils 178mm deep.

The means and the implements available for conversion year cultivation are restricted greatly by two factors. The lines of Keyline cultivation cannot be followed satisfactorily by mouldboards or disc ploughs, nor are those implements suitable for the deeper cultivation which must keep the subsoil under the cultivated surface soil. They also produce the destructive even-bottom cultivation.

They can both be made to do the surface cultivation reasonably well, while another implement of the tine type, with wider spaced rows than the usual farm implements, could complete the deeper cultivation immediately following. Some tine shapes will keep the subsoil down.

Mouldboard ploughs, with the boards removed, give a satisfactory cultivation, if the final deep run is done with some shears removed to keep the furrows wider apart.

Scarifiers or tillers both give a satisfactory surface cultivation to 102mm, but the tine spacing and design render them unsatisfactory for the final full depth run.

Rippers will follow the lines of Keyline cultivation for the final working. It is unnecessary in surface cultivation to do this. The resulting cultivation is satisfactory but the cost with any rigid implement is much higher than it need be.

The Yeomans chisel-type plough is the ideal implement for conversion year cultivation. The following details of Keyline absorption-fertility cultivation both for conversion year and the cultivations in following years, are given for use with this implement.

The standard shank row spacings of the Yeomans plough are 305mm apart, double the spacing of other farm cultivating implements. The Yeomans plough is equipped with tines, spikes or chisels 51mm wide, which are set at 305mm row spacings.

With a suitably powered matched tractor and Yeomans, set the plough's depth to enable the tractor to run without laboring at a good speed. Eight kilometres an hour is recommended if the surface is suitable for that speed. When stumps are encountered, reduce the speed to 5km/h. The "first working" should be 76 to 114mm deep. Large clods may result from a first cultivation which is too deep and could necessitate some special extra work to break them down. Plough three or four parallel runs completely around the area, marking out the area for cultivation clearly.

The paddock area is then "cut out" on this first run by ploughing backwards and forwards, turning in the series of parallel runs first made without necessarily reducing speed on the turns. The Yeomans will follow as fast as the tractor can turn.

Plough a second complete run immediately at a long angle to the first with the plough now set deeper and travelling at the same speed. It is more economical usually to regulate the increased depth to suit the speed and not the speed to suit the depth.

This second plough will sometimes give a suitable surface "break-out" and the necessary depth of 178 to 203mm. If that is so, the second cultivation run will follow the Keyline cultivation principle of Chapter 2.

Usually three fast cultivation runs using 57mm chisels at 305mm spacing are necessary for perfect conversion year cultivation in poor compacted soils. In that case the depth of the last run is set at 178 or 203mm, as already discussed.

That simple Keyline conversion year cultivation will start a cycle of soil fertility which can be carried forward to greater soil improvement and will produce a better than usual crop at the same time. It will also be effective in holding the soil against erosion.

The fast and low cost cultivation will start to improve soil immediately adequate rainfall is supplied. The natural processes of decay will, at once, go into action.

Poor heavy soil, that is soil low in humus content, should be watched closely after heavy rain against a possible surface sealing. If that is apparent the area is given a one-run Keyline cultivation immediately the soil is sufficiently dry. The soil will improve only with adequate air. The first year is one of destiny for the soil.

If a crop has already been sown, it is still often advisable to aerate the sealed surface soil when it is dry enough by the one-run cultivation. The spikes should be spaced at 610mm intervals for the aeration cultivation.

The health of the soil, the progressive development of structure, fertility and soil depth, is of infinitely greater importance to the farmer than any one crop. Such outlook will, however, result in better crops all the time. Even a crop newly out of the ground and partly destroyed by a cultivation to aerate the sealed soil will usually yield better for the treatment.

Conversion year cultivation of poor soil is completed by no more than three fast workings, each becoming progressively deeper. The last working, which is 178 or 203 mm deep, is the only one which follows the Keyline cultivation principle. Spikes are 51mm wide and the spacings between the rows are 305mm.

The increased moisture of conversion cultivation will continue decay processes longer and thus obviate one of the difficulties of stubble mulch farming with disc implements, that of having too little moisture available for rapid and continuous stubble decomposition.

The changes which will take place in soil which has been converted to Keyline absorption-fertility should be watched by the farmer. Only absence of rain will restrict the working of the yeast-like process of soil development.

Examination of the underneath cultivation by removing a square metre of the ploughed soil will disclose the deeper chisel final furrows which knit the soil to the earth.

Make an examination a few weeks after the first good rain has fallen — see the change — smell the soil.

Again when a crop is well grown — examine the deeper broken subsoil — note its further changed condition. Fertility development in the surface millimetres will be apparent and the deeper broken subsoil will be changing into good soil.

When the crop is stripped, examine the condition of the subsoil again to get a cue to the depth of cultivation to be followed for the next crop. If a change from the subsoil to a soil is definite, second year work should be a little deeper. The increase should be a 25mm or so at most. The broken subsoil is to be converted to soil, a little each year — progressively.

In subsequent years, following a successful Keyline conversion-year, a single run on the Keyline cultivation will complete the ploughing. Now spikes or chisels with weed knives attached are used. The weed knives work 76mm below the surface, mixing growth and trash correctly into the soil for rapid decay. At the same time the chisels work at the full cultivation depth, properly aerating the whole body of soil.

The weed knives, which are adjustable in relation to the 51mm, chisel depth, permit a progressively deeper year-by-year cultivation, with the knives working at a fixed depth below the surface. The uneven furrowed type bottom and the "completeness" of all the soil is preserved.

The rate at which beneficial decay will take place in the soil will vary with soils and climatic conditions. The rate of decay accelerates as a positive new soil fertility develops. Decay of the incorporated vegetation of conversion year cultivation will be rapid given sufficient moisture. Decay in subsequent years will be much faster as the active life in the soil has been built up enormously as a result of this conversion to absorption-fertility.

For a short time decay does tend to rob growth of some of its requirements. Both decay and growth require among other elements, moisture, air and nitrogen. A crop sown immediately in conversion cultivated land may first grow weak and yellowed from the lack of nitrogen which has been absorbed temporarily in the processes of decay. With adequate moisture, air and heat, nitrogen will be available to the crop in a few weeks. The crop will respond with a rapid growth of healthy green foliage.

A rapid fertility gain and almost weedless farming on this crop land can be secured by cultivating immediately the crop has been stripped and again each time a growth of grass and weeds reaches its "full green" stage before the weeds' seeding. The use of the chisel and weed knives combination tends to germinate all seeds together, while "soil turning" methods of cultivation do not. The "soil turning" implements bury some seeds in a dry layer of vegetation, which prevents their germination until a later cultivation, thereby assisting the continuance of weed growth.

Weedless farming may disclose that the present row spacing of seeders, which are close together to enable crops to partly choke weed growth, is too close for best yields.

There is a growing well-informed body of opinion among practical Australian farmers that wider apart seeder row-spacings will give better grain yields when weeds are not a factor.

The sowing of seed into conversion-year cultivation requires some little changes from the orthodox habits.

It is of particular significance that sowing be done in such a way that the new condition of the soil is preserved as much as possible. If an ordinary grain combine is used, the cultivating tines, both the front and rear rows are removed permanently, use being made only of the two planting rows. The soil will continue to be in a suitable condition for the rapid absorption of rainfall. The use of all the tine rows on a combine may so "fine-up" the soil that it will cause puddling and washing. The use of the combine with all the cultivating tines removed will permit rougher and trashier ground to be seeded.

Planting depth for grain will vary also, but generally seed should be planted into the moisture zone and not shallow-sown with complete dependence on later rain for germination.

Soils of good structure and fertility may be cultivated directly with the chisel and weed knife combination. If surface cultivation ever appears to be too fine, use the chisels only.

Cultivation of soil in very low rainfall areas should be accomplished by a shallow surface cultivation followed by a final Keyline cultivation with the chisels 0.6m apart. An overall cultivation that is too deep on these soils will tend to lower the moisture zone too much for best yields. As the fertility of the soil increases humus will protect the moisture and hold it at a more consistent level.

Once a normal rainfall season follows, or good rain out of season has fallen on Keyline converted land, the moisture horizon will be more dependable. Continued year by year, Keyline absorption-fertility cultivation will keep adequate crop moisture available for longer and longer periods into dryer times. No doubt later on the "Keyland", one good season's rain will produce two years good crops.

The low cost and speed of the method of cultivation is apparent. Conversion-year cultivation will usually cost less in time and money than extraction-fertility methods. Following conversion-year, costs are about one-third only of old cultivation habits.

Deep fertile soil, then, is built up for crop land first by conversion year cultivation with an increase in the depth of the chisel penetration each subsequent year. The weed knives work about 76mm below the surface. In from three to five years soil depth to the limit of the Yeomans 406mm is formed.

When that depth is reached, another "wave of fertility" may be induced in the soil by reducing the penetration depth of the chisel back to about 229mm and adjusting the weed knives to work deeper under the surface.

Instead of 76mm, as used in the first cycle of increasing depths, they are used 127 or 152mm below the surface. For the following two to four years increase the working depth of the chisels 25 to 76mm each year, but adjust the weed knives to keep them 127 to 229mm below the surface.

The effect of the second series of increasing depth cultivations and increased weed knife depth is expected to add a greater depth of intensely fertile soil By incorporating or mixing vegetation into a greater depth of top soil that should be achieved. The soil should now be in a condition to "take" the somewhat deeper mixing of vegetation, whereas in the first years it would have been largely lost as a fertility gain. At the end of the second cycle — originally poor soils in reasonable rainfall areas do rival the most fertile soils left on the face of the earth.

CHAPTER SIX

Soil improvement on pasture lands

ON GRAZING PROPERTIES GENERALLY, compaction of the soil has developed from the tramping of stock. That tends to limit the pore space and the free movement of oxygen in the soil. These soils change as distinct zones are formed by compacted horizons below the shallow grass root depth. The natural movements throughout the whole of the soil become more restricted, less deep mineral material finds its way to the topsoil to replenish it, and the soil gradually becomes impoverished in both humus and minerals. An unnatural division of the soil into layers is made. Only the shallow top soil, with its rapidly diminishing minerals, is available.

Good grass species tend to run out, as the whole pasture deteriorates; less rain is absorbed; soil losses may occur; valleys become too moist and sour or they erode; hills become dryer and less productive.

Such pasture now needs two things it has lost and which can be supplied by proper cultivation, enabling its processes to be stimulated again. They are air and water; or simply absorption capacity which will enable the soil to absorb and hold the rain that falls on it. Although the soil cannot be put back in perfect condition in one cultivation, it can be progressively improved to a condition usually better than it was originally.

Single working Keyline cultivation with a depth of penetration just through the top soil into the compacted zone is a logical first means to supply both the air and moisture required. Excellent results will follow that work completed in the autumn. Another suitable time is probably a few weeks before each locality's best rainfall season.

Spike or chisel furrows 305mm apart and at the depth previously suggested, break or crack the continuous horizon of compacted material which now divides the full depth of this soil. With aeration and quick moisture penetration the wholeness or completeness of all the soil's depth is brought back progressively. The decay of dead and dying root growth again adds rapidly to soil fertility by the formation of humus below the pasture.

Some pasture grass is destroyed in the process by being uprooted, and more pasture becomes temporarily unavailable to stock by being partly clod and sod covered. Within a very short period a much improved pasture, both in quality and quantity, is again available. The soil is coming to life again.

It may be appropriate at such time also to introduce new species of clover or grasses to assist the development further. The use of lime or fertiliser is often of considerable advantage in starting a new cycle of fertility in the poor soils.

It is often highly profitable to conduct a two-or three-year plan for the improvement of a very poor soil paddock.

> *First Year.* — One Keyline cultivation working about 114mm deep with spike spacing 305mm apart is given in the autumn.
> *Second Year.* — Another Keyline cultivation working 127 to 179mm deep with spike spacing 610mm apart.
> *Third Year.* — A third Keyline cultivation working 178 to 254mm deep with spike spacing 0.9mm apart.

Stock is moved off the area immediately before each Keyline cultivation working and not returned until some weeks after the first rain has fallen on the area.

The clods quickly become improved in structure and are partly distributed by the stock over the surface, thus forming a valuable topdressing to promote greater absorption, decay and fertility. Careful stocking of that treated pasture can make it still more effective. Soils so treated are in a perfect condition, especially if frosts have worked on the clods, for rapid response to all other means of increasing soil fertility and yield.

Rotational grazing, strip grazing and smear harrow treatment, by greatly increasing the effectiveness of the use of the fertility potential of the animal droppings, are outstanding in their fertility effect on this treated soil.

By such means poor shallow soils will not only become more fertile, but will be converted profitably into areas of considerably increased soil depth.

If the depth of fertile soil is doubled, the profit margin is increased many times. The aid of progressive development by progressive increase in penetration depths for maximum absorption-fertility is of outstanding importance.

The drastic ripping or subsoiling of pastures on the poorer thin soils, while probably increasing first year yield, will all too often be disappointing in yield for following years. Deep ripping with rigid implements is very costly and throws up clods which are too big. Heavy soil will not remain open to that depth but, will reseal with the first good rain. There is no profit in taking depth which cannot be held. The topsoil fertility will fail to produce a rapid soil change in the subsoil if it is given too much depth of subsoil to "convert". Again consider the topsoil as a yeast and do not subject it to too great a dilution as may take place in the case of overall deep ripping or subsoiling.

The present methods of subsoiling crop land, where deep subsoilers rip the soil to 610mm deep, and surface cultivating implements follow, is wrong. The fine surface cultivation of deep subsoiling largely offsets the benefits of the moisture and air absorption capacity of the subsoiling. All the benefits of sub-soiling, without its usual disadvantages and high cost, are obtained in the final deeper run of Keyline cultivation. Extra depth can be obtained by increasing the cultivating row spaces.

The object of Keyline cultivating below the soil into the subsoil is always the improvement of soil fertility and the conversion of this subsoil into more fertile soil. It can be done most profitably and economically only as a progressive process.

Compacted soils of all types have lost the natural fertility potential available to all soils of good structure. The continuous decay and humus formation from the considerable amount of grass roots material which dies each year is almost entirely wasted.

Poor compacted pasture land usually has available to it every ingredient for a rapid fertility increase except oxygen and water, and those two are prevented from being used fully.

Minerals of all necessary kinds are usually only millimetres below the poor pasture. The urine and dung of the animals are available in sufficient quantity. Materials for aerobic decay and humus formation exist in the grass roots, all of which have not been completely lost.

One low cost fast run with spikes or chisels on the Keyline principle makes available all the ingredients for a new fertility. With a few weeks after rain on that cultivation, the return of life to the soil and pasture can be seen in the rapidly changing structure of the soil.

Whenever pasture land shows sign of surface sealing or compaction it should be treated in that way. If and when the second cultivation is required it is made deeper. The same high speed and low cost is obtained by increasing the spike or chisel row spacings. Actual soil depth is increased that way.

As soil becomes more and more fertile, less and less aeration by cultivation is necessary. Reasonably well managed highly fertile soil will look after itself. It will absorb all the available factors of fertility and aerate itself. It will preserve its own "life", including the beneficial earthworms.

Fertile soil and pasture absorb moisture rapidly, store it deeply and the soil aerates itself.

Other plant nutrients as well as oxygen and water reach the earth in the rainfall. They are largely absorbed into the soil and held if the soil is properly treated.

When poor pasture land is to be completely cultivated to kill all growth for the replanting of a new pasture, it is treated as described for conversion year cultivation.

Grass seeds are sown into the cultivation with outstanding results, by seeding with an ordinary grain combine with the cultivating tines removed. For even sowing and better germination, a flow medium of some kind mixed with the grass seed is a great advantage. Sow into the moisture zone some time after rain has fallen.

If the soil is of poor structure — low in humus — watch it against possible surface sealing after heavy rain. If it seals give it one working when it is dry enough. Follow the Keyline cultivation with spike spacings 610mm apart.

If pasture tends to run out something is definitely wrong. Apart from over-stocking or indifferent stocking management, the cause will be moisture wastage — shortage of oxygen — or both. If pasture land is assisted by correct cultivation to absorb moisture and air it will continue to improve in fertility and productiveness and will not run out.

Today most pastures tend to deteriorate, and those declining pastures are ploughed up, a crop or two taken and re-sown again to grass and legumes. The poorest pasture paddock is usually selected to be used in this way.

If crops are to be taken on pasture land only good pastures should be used. Any farmer would be reluctant now to take this course, but if all his pastures were good, he may select his best pasture paddock for cropping. Any three-year-old pasture should be good, and improving. The newly sown pasture will probably be the lowest yielding, but will be improving rapidly with the soil fertility. The farmer will select his best soil and pasture for his crops and so allow time for his newer and poorer pasture to improve with the soil before they in turn come up for cropping.

Some pastures may need Keyline cultivation for fertility by absorption each year for two years, and need the treatment again in three years, then five or more years later.

As both the soil and pasture improve, better grasses may be introduced with any Keyline cultivation. An improving soil will more truly indicate its requirements in minerals or trace elements — should they be necessary — than a soil which is being forced to yield by one or more of the popular methods of extraction-fertility.

There are many methods and techniques for pasture improvement, some good, some very bad.

Pasture improvement can be obtained — temporarily at least — by more efficient methods of extracting the remaining fertility of the soil. It can be secured properly and permanently only by methods which primarily improve soil structure, fertility and depth. It is wrong to use chemical fertilisers only to improve pastures. A fertiliser, if used, should be used in such a way as directly to improve the soil. This improved soil will give an improved pasture, thus starting a cycle of soil fertility, permanently improving pastures.

If fertilisers cannot be used on soil apparently requiring them to assist directly in "triggering-off" the new cycle of soil fertility, the soil is much better off without the fertiliser.

Soils that do require the use of fertiliser also usually need, and much more urgently, the application of the principles of absorption-fertility. If the soil is very low in humus, the first full green growth on such soil should be ploughed into it. That will start the cycles of fertility and increasing yields. Fertilise to improve soil, and depend on improved soil only for increased yields.

The recent enthusiasm for pasture improvement in Australia has unfortunately emphasised the wrong word. "Soil improvement" is the only real basis for long-term pasture improvement.

It is more than likely, indeed almost certain, that the introduction of new grasses and fertiliser to increase rapidly the stock-carrying capacity of poor soil, is providing the farmer with another method of extracting the fertility of the soil. The soil must always be considered first. Increase absorption, manufacture humus under the pasture, improve the structure of the soil, increase soil "life", then the improved grasses will readily assist in the full development of soil fertility and produce abundant pastures.

It is fully recognised, however, that some methods and techniques have produced outstanding pastures.

Disc implements have on occasions been used exclusively, and have improved soil and pasture on soil which had lost its condition and some fertility. The shallow disc ploughing into the soil of crops of weeds and later sowing pasture grasses by the methods of broadcasting or "direct-drop" and then harrowing, may give an outstanding pasture for a few years.

By improving top soil fertility, actual improvement of soil depth may take place very slowly, but the pastures tend to "run-out".

Such pasture treated by the Keyline method for soil and pasture improvement will produce rapid and permanent soil and pasture improvement.

Very fertile soils on occasions require Keyline cultivation. After big floods recede from farming and grazing land there is usually striking evidence of the damage caused to the soil by waterlogging. The soil has been killed by too much water. It is literally "dying for air". Pastures which grow out of this soil are not healthy stock food, although the grass may be growing well. It is the type of food suitable for the hordes of pests which feed on the products of infertile or "sour" soils. The pests locate that food and devour it as they breed in countless millions. They may "foul" the soil to such an extent that stock will not graze what may remain. With the infestation, weeds often grow in profusion.

The soil needs Keyline aeration cultivation immediately it is dry enough. The "sickness" is then cured and the soil will be almost immune to those pests. A fast working of the land with tines spaced at 305 or 610mm apart at a depth of 102 to 127mm is all that may be necessary to bring the soil back to a healthy state. Deeper cultivation depth on the wider spacing could be considered.

Disc implements and mouldboard ploughs are not recommended because they are unsuitable for following the lines of Keyline cultivation. They do not promote rapid soil improvement and are incapable of the correct deeper cultivation.

Mixed growths of vine and rough grass may be given one shallow run with a disc implement and immediately followed with the spiked implements. Keyline cultivation must always be followed.

An outstanding pair of implements for soil improvement particularly where the growth is heavy and matted are the mulch mower and the Yeomans plough. Any good flail mower will do the job efficiently.

The mulch mower can be used also to the great benefit of the soil any time pasture growth is high and not required for immediate stocking or fodder conservation.

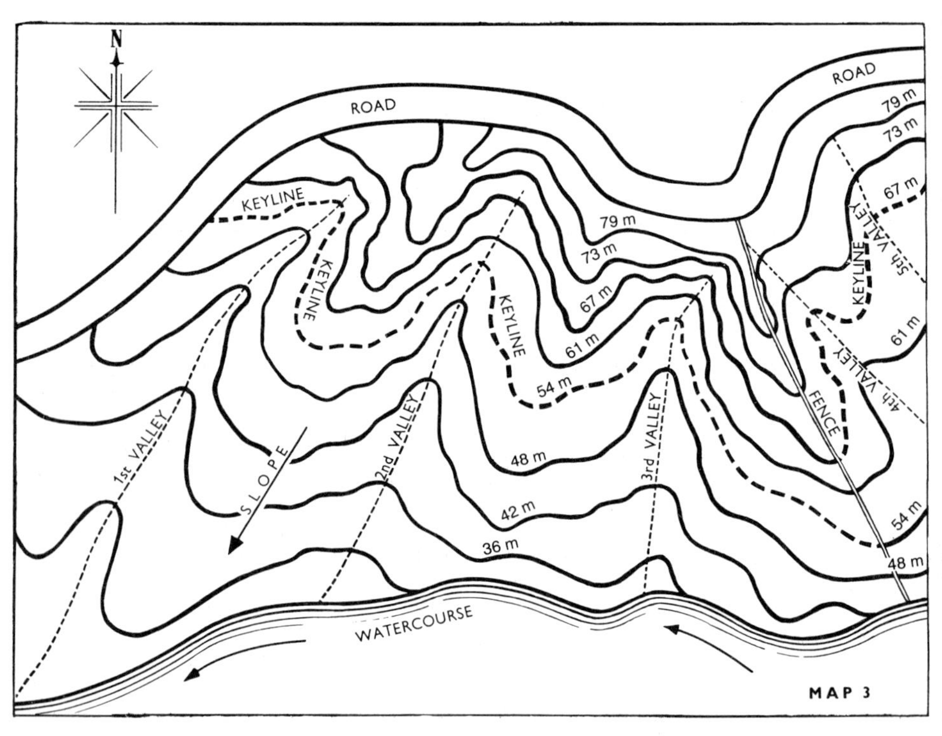

N
ROAD
ROAD
79 m
73 m
67 m
KEYLINE
KEYLINE
KEYLINE
KEYLINE
5th VALLEY
79 m
73 m
67 m
61 m
61 m
54 m
4th VALLEY
FENCE
1st VALLEY
SLOPE
2nd VALLEY
3rd VALLEY
48 m
42 m
36 m
54 m
48 m
WATERCOURSE
MAP 3

CHAPTER SEVEN

Common Keylines and Keyline land units

FROM THE LIMITED APPLICATION of the Keyline of one valley as illustrated on Maps 1 and 2 and discussed in Chapters 1 and 2, we now consider the next step — the extensions of the Keyline.

Each valley will have its Keypoint and Keyline. Where, by the extensions of the Keyline levels — either on a true contour or with a slight fall — the one Keyline serves two or more valleys, this line becomes a "common Keyline". It is simply one line of levels that forms the Keyline of each valley it crosses.

Map 3 illustrates an area of steep country with five major valleys draining towards a rocky creek. (See page 50).

An examination of the first valley indicates that the 54m contour — the broken line — is the Keyline of this valley. The same contour also serves as the Keyline of the second and third valleys but crosses the fourth valley in a location obviously not the Keyline of that valley. The 54m contour is the common Keyline of the first, second and third valleys, while the 67m contour line is the common Keyline of the fourth and fifth valleys.

For purposes of cultivation and development those two common Keylines control two separate areas. A fenceline up the centre of the ridge between the third and fourth valleys divides the areas according to common Keylines. The two sections are Keyline areas, or complete Keyline cultivation and development units. They include the areas both above and below the Keyline.

A Keyline area, then, is an area controlled by a Keyline or a common Keyline and may include any number of valley areas. The Keyline areas of Map 3 may be again subdivided into any number of paddocks.

Conversion-year cultivation in the case of crop land or Keyline soil development for pasture improvement is first completed in the area above the Keyline and parallels the Keyline up the slope of the land.

If the Keyline is not to form a gently falling water race — it often does, see Water Storage, Chapter 8 — some other means of permanently marking and preserving the Keyline is necessary.

A row of stakes first marks the Keyline. Leave a narrow strip unploughed on each side of the Keyline stakes. On it, brush or trees will grow along the line during the time the area is closed to stock for cropping. The line need only be a metre or so wide and it will serve as a permanent marker for the Keyline.

Without the tree growth on the unploughed Keyline strip, a marker can be satisfactorily preserved by carefully following the lines of the previous cultivation. Another means of permanently marking the Keyline is to use it as a farm roadway.

CHAPTER EIGHT

Water storage

IT HAS ALREADY BEEN STATED that the greatest available water storage capacity exists in the soil itself. The association of Keyline cultivation and such water storage capacity has already been explained.

If all rain which falls on crop and pasture land could be absorbed into the soil, there would still be areas remaining that do shed most of the rain which falls on them. Farm roads and yards, the homestead and other farm buildings and sheds, and often main roads, shed considerable quantities of rain. Conservation of that water for farm use is of the utmost importance.

Whether a farmer realises it or not, he is dealing with forces which need the full use of engineering planning. A sudden storm may send 100,000 tonnes or 500,000 tonnes of water on to a 405ha area in an hour or two. That huge weight of water can be controlled and conserved by the farmer to the great benefit of the land and himself, or it can run largely to waste, leaving a trail of destruction in its path.

Levels are important factors in any water control and conservation project. They need to be used to advantage by the farmer. Contours and other level considerations are basic land engineering factors. The farmer must know how and when to use them.

The application of Keyline methods requires very little levelling work, but those levels it does require are of great importance.

On undulating country, dams can usually be located which will enable the farmer to enlist the forces of gravity to provide him with water under pressure. That will give him a better farm, easier work and higher yields. Other things being equal, the value to a farmer of conserved water is in direct proportion to the height of the storage. The dams of potentially greatest value are those in his high country.

The Keylines, by crossing the valleys at their first main point of slope flattening, will invariably position the highest suitable valley storage area for water.

In any plan of general land development, the control of water is one of the first considerations. At the same time, it is to be kept in mind that Keyline absorption-fertility is going to reduce runoff water very considerably. It may even completely stop runoff water from farm and pasture paddocks, except in the rare, but in present conditions, very dangerous period of general heavy flood rains. With the absorption of what previously would have been heavy runoff, consideration has to be given to conserving water from every available source.

With the Keyline positioning the highest suitable dam sites, it becomes important to locate potential water-shedding areas above the Keyline.

The Keylines have been illustrated as contour lines in the discussion on cultivation for the sake of simplicity.

For purposes in connection with the conservation of water in the Keyline dam, the Keyline itself is a gently falling line to form a drain or water race to carry water to the dam. The use of the Keyline, which is now a drain, is still fully effective as a guide for Keyline cultivation.

It is usually convenient and good practice from most other viewpoints, to locate the homestead and all farm buildings and the yards and their attendant roads in the higher country. From the point of view of full Keyline develop-

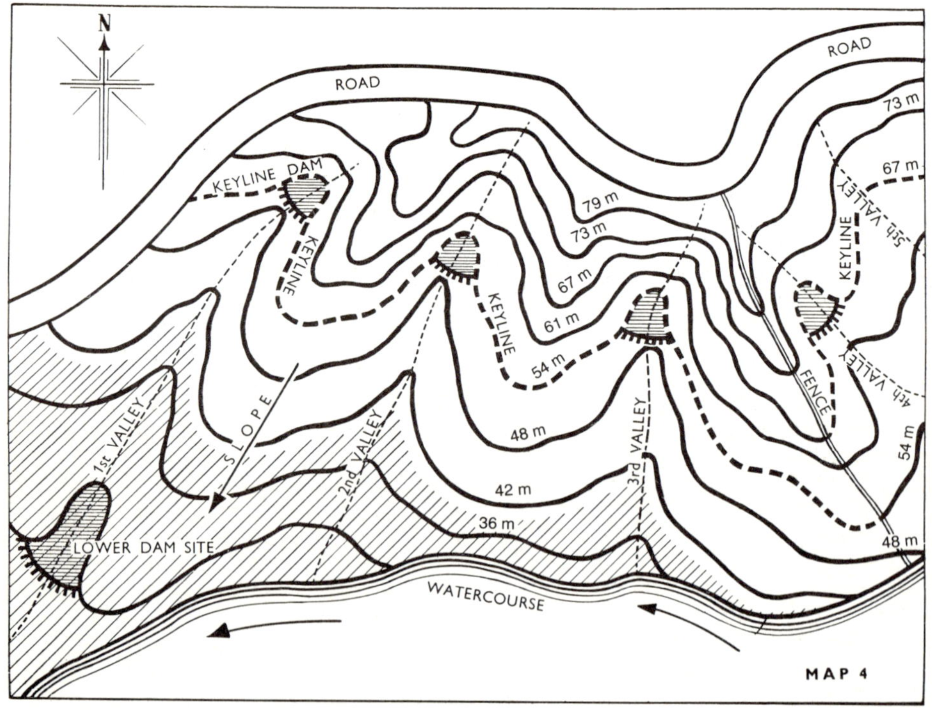

ment, it becomes a part of planning to do so, to secure abundant runoff water to fill the Keyline dams from those sources.

Wherever it is possible and practical, dams are constructed on the Keyline in the valleys, and the Keyline itself is pegged and constructed as a gently falling drain to carry water to the Keyline dams.

Keyline dams are constructed with a pipeline through the wall or through the floor to one side of the centre line of the valley, so that the full gravity pressure of the conserved water is available for spray irrigation and other farm purposes.

Where areas of land exist that are 15m or more vertically lower than the Keyline, the water from the Keyline dam will supply effective pressure for irrigation without pumping. That "line of effective water pressure" suitably forms the top boundary for the irrigation paddocks. A 102mm pipe through the wall, controlled by a 102mm gate valve, in those circumstances will control gravity pressure which, often from a single dam, will effectively work a comprehensive spray irrigation and stock-watering system.

With the use of a 102mm pipeline, the vertical drop from the water level to a nearby irrigation area multiplied by .4 will give the approximate pounds pressure available in the spray line. A vertical fall of 15m multiplied by .4 gives 138kPa pressure, which is suitable for most types of spray lines. As the spray line is moved downhill a little on each "move", there is, of course, an increase in available pressure.

Referring to Map 4, which exhibits the same land area as Map 3, the Keyline crossings of the valleys are to be considered as possible dam sites. The sites marked in four of the valleys could be considered good dam sites. The site of the Keyline crossing of the fifth valley is not as suitable as the others.

The most valuable water storage site for a Keyline dam is located in the first valley, as that site has the greatest area of land below it which is suitable for irrigation by gravity sprays. That fact indicates a rule or general formula for determining the direction of flow of the Keyline when it is formed by a drain. If the creek or drainage line below a series of valleys — as in Map 4 — has a general fall greater than 1.5m per 305m — the fall recommended for the Keyline drain — the direction of the Keyline fall follows that of the creek. When the creek has a flatter fall than required by the Keyline drain, the drain falls in the direction opposite that of the creek. That is illustrated by the shaded area on the map.

The construction of a Keyline dam will often cost considerably less than a pump and engine installed for spray irrigation. The Keyline dam, its pipe and valve outlet, will work the same sprays without pumping cost.

That low cost water is used in the general program of progressive soil development, and higher yields will be incidental and automatic to the Keyline absorption-fertility program.

The following construction comments should be considered.

Most undulating country is suitable for dam construction if correct preparation and compaction of the material in the wall is secured. Fine clay, which is usually considered the best material for dam bank construction, has its own particular problem. That material in the wall of the dam will tend to "jell-up" below the waterline to such an extent that the weight of the wall above the wet unstable material may squeeze the material outwards from the wall, thus causing a central subsidence of the wall which extends down below the waterline. It would result in the water overflowing at that point and would completely destroy the bank.

In shale country a mixture of shale and clay will give the best possible material for bank construction.

Before laying-in a dam bank, the foundation area of the bank must be treated first according to the type of country. In shale country it is necessary to remove only the darker topsoil material to one side — it can be used later to cover the bank to obtain a quick growth of grass. The cleared area is then ripped before the wall filling material is placed on it. The material for the wall should be placed on in layers of from 152 to 305mm thick, so that suitable compaction of the soil takes place during construction. Bulldozers will give sufficient compaction usually without the need of other special compacting implements.

The back of the wall of the dam, that is the side away from the water, should not be specially compacted. If water seeps through the compacted front of the wall into the centre, it must be allowed to get out through the back of the wall, otherwise it may build up hydrostatic pressure inside the wall. That could destroy the wall by forcing or breaking the material from the back of it.

Clean water seeping through a dam wall is usually quite safe, but seepage discolored by the wall material should be considered a danger to the wall itself. Raking or harrowing of the side of the wall in the water of the dam is usually the best means of sealing that type of seepage.

In the construction of that type of dam by bulldozers, the excavation of the sides of the dam, if the land will stand firm, should be made on as steep a slope as the implement will dig. The water-side of the wall, as formed by the action of the bulldozer pushing the material upwards, should be flatter than the excavated sides. Usually the limitations of the implement to push material up the slope of the wall will form a wall of suitable slope.

The laying of a pipeline through the wall of the dam, or through the earth below the wall of the dam, requires special attention.

The danger to be avoided lies in the fact that water will tend to flow along the outside of the smooth pipe, creating an ever-widening and larger hole, which may eventually let all the water go and so destroy the wall.

[Dam building in detail is covered in the second half of this book.]

The high contour dam is the highest dam of the Keyline plan. It is located in the areas above the Keylines.

Gently sloping country usually exists above the steeper slopes which lie above the Keyline. The valley heads will actually start at the low edge of this flatter country where the steep slope country begins. The high contour dam is constructed there. The area selected for the dam site can be the side of a hill or ridge. A slope as steep as 1 in 10 is suitable.

The race or drain to transport water to fill the dam is located above the valley heads. It also serves to further protect the valleys by preventing any flow into them. The drain requires a fall of about 145m per 305m. The site of the drain and dam must be studied and planned together.

A sketch and cross section of a high contour dam built on the steep slope mentioned is illustrated on page 64. Each cubic metre of earth moved conserves two cubic metres of water. That ratio is not as favorable as that in the construction of Keyline and other valley dams which may be around the ratio of six of water to one of excavated material. However, the value of the conserved water in the high contour dam more than warrants its construction where the topography is suitable.

The high contour dam may be constructed anywhere along a ridge where a suitable slope exists and where runoff water can be brought to the dam by a drain from one or both directions.

Because of those circumstances, the dam is usually long and narrow and always along the contour.

A bulldozer is used for construction and the earth is moved from the topside straight across the dam to form the wall. In that way the haul is lessened and the cost of earthmoving is in direct proportion to the distances the earth is moved, so the distance is kept to a minimum.

The drain to fill the dam is located and pegged when the dam is marked out. The construction of the dam is completed before the drain is built. There is then no danger whatever of losing from heavy rains any part of the dam during its construction. The back wall of the dam is constructed first. Then the 102mm pipe outlets are laid at one, or both, ends. After that, the end walls are closed and the drain made.

A spillway is not constructed, because surplus water is allowed to overflow from the drain at some distance from the dam when it is full. It is necessary only to see that the overflow does not occur at the same place more than once during the first year or two, so that no water wash is started. Once the drain is grassed, blocks can be made at any suitable place in the drain to overflow the water there.

Water transporting drains can become less effective, or sometimes completely ineffective, by becoming overgrown with vegetation. The best means of controlling growth is by seeding the drain to good grass species and manuring the drain heavier than the adjacent pasture. That encourages stock to graze the drain area more closely than the rest of the paddock. It is also advantageous to mow regularly the long excavated slope of the drain so that the water transporting capacity of the drain is unimpaired.

If a road is to traverse the area of the drain it can be placed parallel to and above the drain. The water run-off from the road is caught by the drain and conserved.

The Keyline dam, constructed on the Keyline, and the high contour dam, above the Keyline, are the two highest dams used in Keyline planning. For that reason they are the most important dams of all water-conservation schemes.

The water conserved in those dams is available under pressure for instant use. It is the lowest cost irrigation of all conserved water and is, therefore, used when the first dry spell makes its use profitable and advisable. No dam should ever be completely emptied except for reconstruction or enlargement. A metre or so of water is always left in the dams, and it will go a long way toward protecting a bank from dangerous dry cracking.

There are many farms which do not have their own Keylines. The development of such farms is mentioned in a later chapter. The conservation of water below the Keyline and on properties of lesser slopes is discussed here.

The first of such dams is called the guideline dam, and is, like the Keyline dam, a valley dam. The wall material is excavated from the area of the valley which will be below water level when the dam is filled. All earlier comments about the Keyline dam, including the pipe outlet, are common to this dam. Its particular location is apparent from the chapter Flatter Lands.

The next dam in Keyline planning has its counterpart in the ordinary valley dam. These are to be seen on farms and grazing properties all over the countryside. The main consideration in locating the usual farm valley dam has been to conserve the greatest amount of water for the earth moved.

HIGH CONTOUR DAM

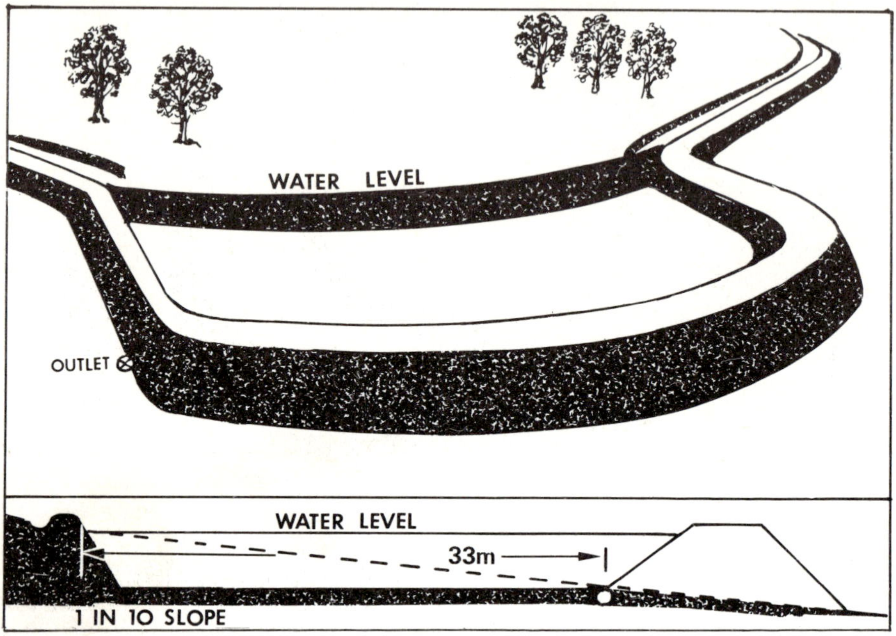

CONTOUR DAM

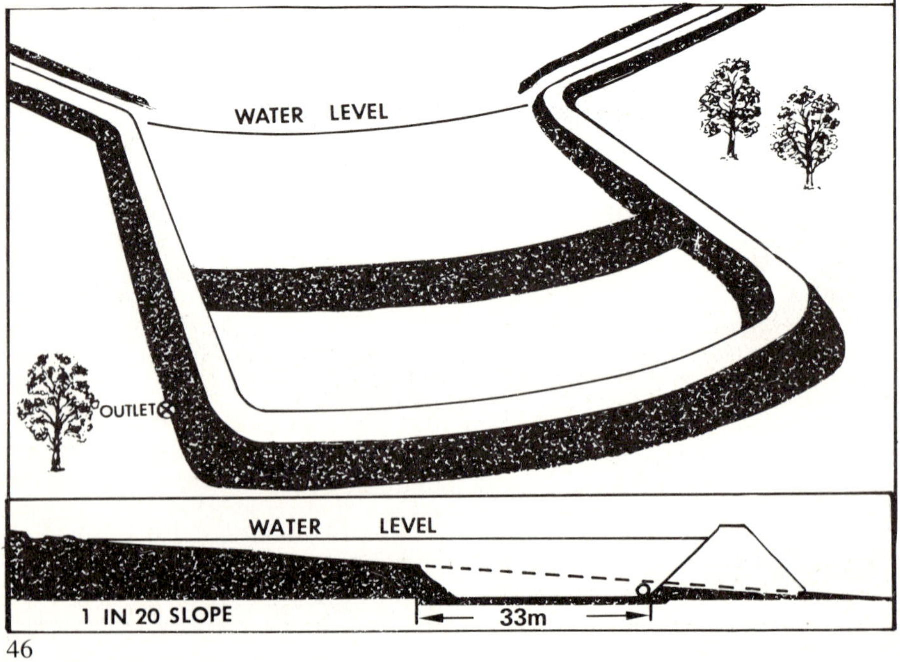

BROKEN RING DAM

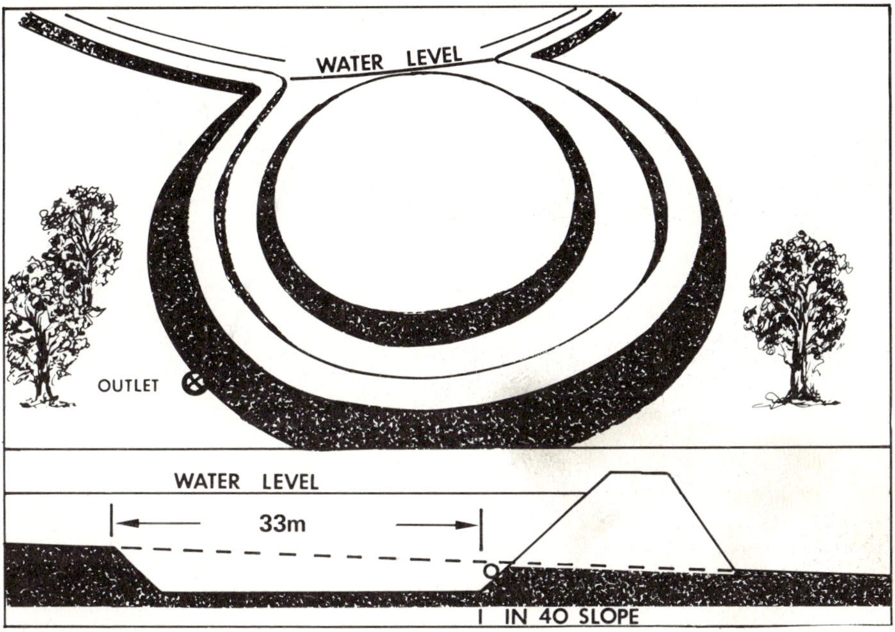

COMPLETE RING DAM

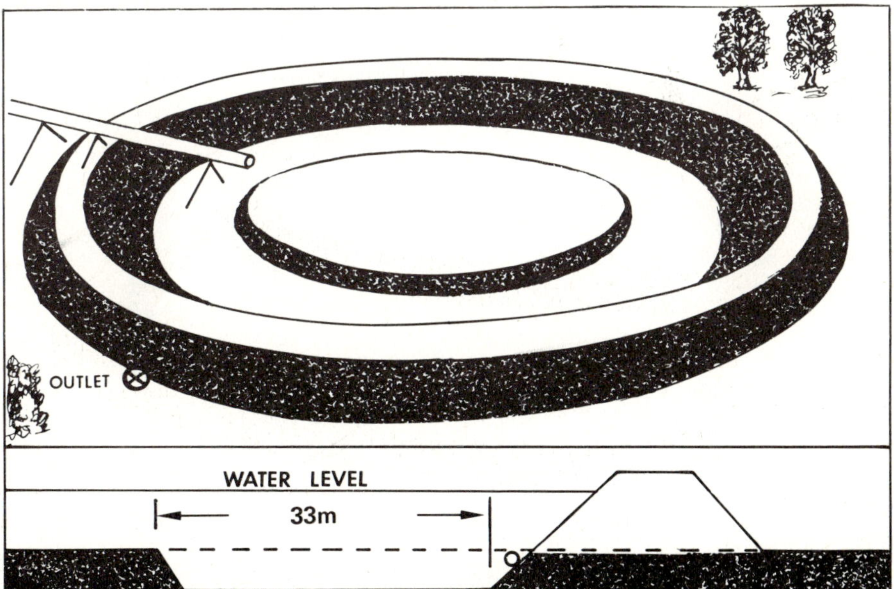

With the absorption into the soil and the conservation in Keyline, high contour and guideline dams, of practically all the rainfall, a large capacity lower dam has to be located where it can be filled despite the other storages. By locating it in a lower valley, such as the site indicated on Map 4, it is in a favorable position to receive the combined seepages from all the higher country. Apart from seepages, it will receive water from very heavy storms and in the periods of general heavy flood rains when most water conservation storages may overflow.

The dams can be made large to act as a buffer or safety against prolonged drought. They should be as deep as practicable, so evaporation losses are reduced. Losses by evaporation are in proportion to the surface area of the water. A dam 1.8m deep could lose all its water in a hot dry year, while a deeper dam would lose only the same depth and have water storage when the other is empty.

The construction of dams by blocking a stream or creek is usually controlled by a government water conservation and irrigation authority. Plans usually need the approval of the authority, which will also often assist with advice on the preparation of the construction plans. Apart from other constructional details, the provision of adequate and safe spillways for overflow is of maximum importance in stream dams.

Contour dams, of which the high contour dam is the one placed in the highest location, can be constructed in almost any type of country to provide low cost large capacity water storage. They are not located in valleys and, as with the high contour dam, require drains to provide the water.

On the land below the Keylines they can be filled from a flowing stream or one that flows intermittently.

The location of a contour dam is decided by first, the means to fill the dam, and second, a suitable area for the use of the conserved water. The water may be used for spray irrigation and other purposes. The main total cross sectional area of the excavation and bank are about the same whether the dam is very large or of medium size.

In the construction of the contour dam a bulldozer is used and earth is moved straight down the slope at a right angle to the contour. The distance of the "haul" is kept to roughly 30m, to provide for the most efficient bulldozer use.

A similar construction to that of the high contour dam is followed. In flatter country the end walls — which are the same length as the width in the high contour dam — become longer. In the high contour dam all water is conserved by holding it in the excavated area by the wall. The contour dam, on the other hand, holds much of its capacity over unexcavated land.

In flatter country, where the contour dam then assumes the shape of a "broken ring", the end walls are turned in toward each other. The water race feeds the water into the dam between the converging end walls.

On still flatter sites it assumes the shape of a "complete ring" and the major storage capacity in larger dams is then over the unexcavated central area.

A pipe outlet is placed through the end wall of the contour dam at the lowest ground level, and water conserved above that height can be released by gravity.

Gravity pressure is used for irrigation if the conserved water is high enough.

The outlet pipe through the wall of such dams can lead directly to a centrifugal pump outside the wall. That maintains the pump under a positive water head, so instantaneous water pumping is available without pump priming.

A complete ring dam should be constructed on a flat area of land below a ridge to which water can be brought by the drain. In deciding the location of the complete ring dam consideration is first given to the filling the dam by flow

DRY AUSTRALIA

Aerial picture shows water everywhere in the dry country near Bourke in Western NSW.

PLATE 1

PLATE 2

Soil change. An early picture on Nevallan. The only earth not changed by the soil improvement program was that adhering to the roots of a fallen tree. All soil changed from yellow or white to the color of the spade full. The soil fertility improvement aspects of Keyline are obvious.

PLATE 3

Preparing to remove an irrigation flag controlling a flow of 1350 m³ (300,000 gallons) an hour. The flag is 2 m (6 feet) square. An aluminium tube holds the flag in the foreground, and a chain with a spike on each end is attached on the other side. In the picture one spike has been removed from the ground and the chain is being pulled to allow a little water to flow under the flag to fill a second flag. This procedure ensures that the rush of water following the removal of this flag will not wash the second flag out of its position. Irrigation rate 2.4 ha an hour.

PLATE 4

Some of the 80 members of a three-day Keyline school near 'high wall' dam on Yobarnie. Depth of water is 10.1 m and it is the highest primary valley dam on the property.

PLATE 5

Yobarnie photographed from the air, after 17 years of Keyline irrigation development. The property covers about 307 ha and 15 full farm irrigation dams are visible in this picture. (Photographed by Douglass Baglin.)

52

PLATE 6

The flood overflow entering the diversion channel at the creek dam on the western boundary of Yobarnie. This is water of number 3 category (see chapter seven) flowing to the land from outside. At that time the five dams of the chain were filled and the lock-pipe through the wall of the dam is open and has reduced the flow in the division channel. The flow from the lock-pipe and the water in the creek below the dam are out of sight below the wall.

WATER CONTROL IN FLOOD RAINS ON YOBARNIE

PLATE 7

These are the second and third of the chain of dams and the diversion channel. The fourth dam is in the primary valley beyond the distant two loops of the channel and the fifth dam is to the right and beyond the distant glimpse of the diversion channel seen at the right top of the picture. See aerial picture of Yobarnie on previous page.

PLATE 8

The open lock-pipe valve of the second dam of this chain of five is opened to return water to the creek. Beyond it is a second creek dam with water flowing from its spillway. This large creek dam is too low down for the use of its water by gravity and it has a permanent pump on it which is used to keep the dams of the two series filled in dry times.

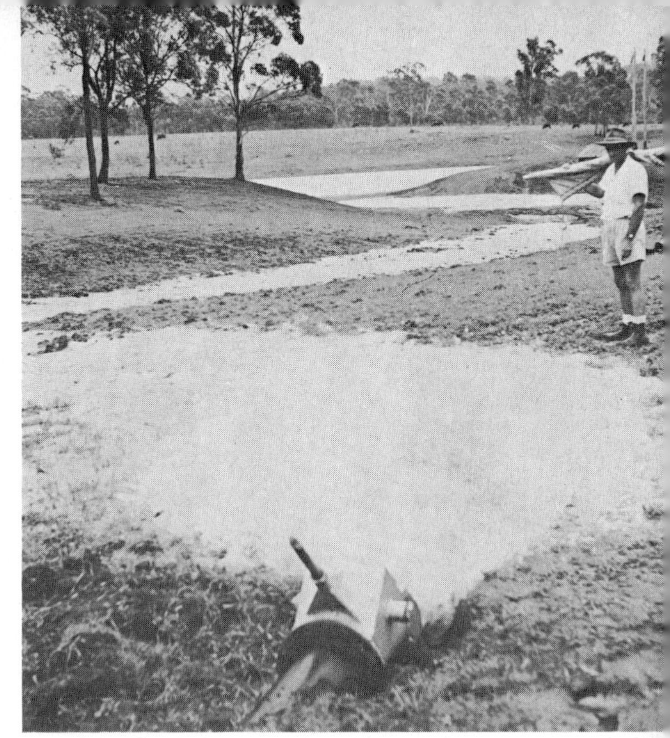

PLATE 9

Water from heavy rain in excess of the capacity of a dam is spilled by an irrigation flag (not in picture) down over pasture land to flow from the property.

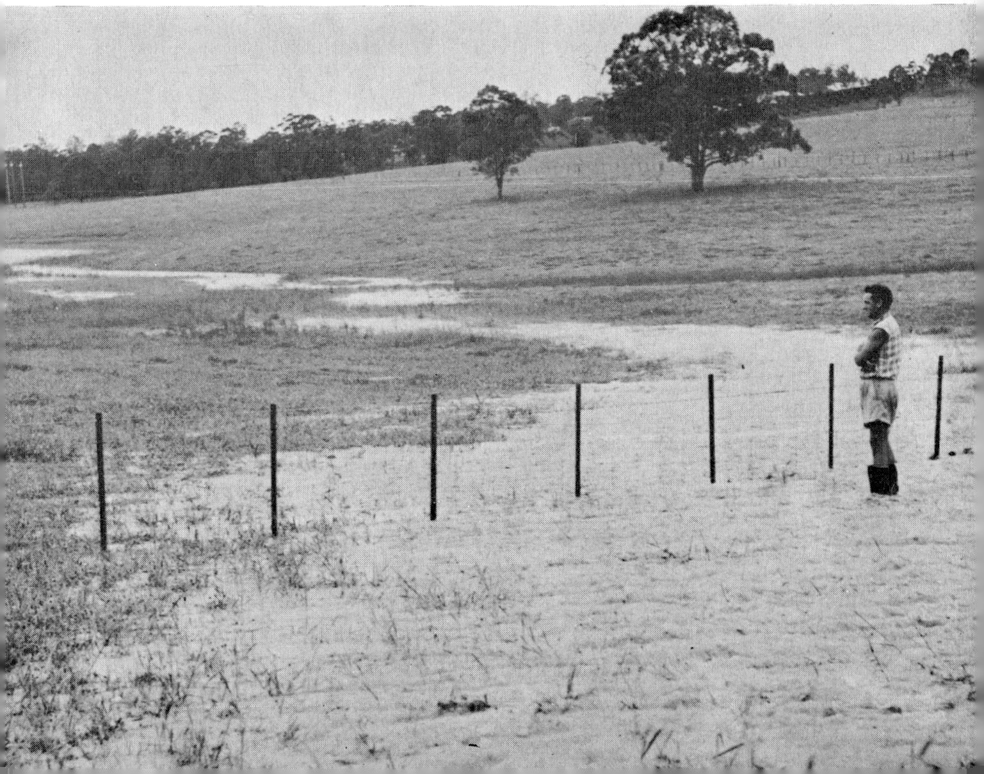

PLATE 10

WATER GATES. Two 148 m water gates, one opened, are for the release of the top 457 mm of the water of a large dam on a grazing property. The dam is maintained filled for a large part of the year by the diversion into it of a fluctuating but constantly flowing stream. The wall of the dam was constructed across a small creek immediately below a large swamp which is now covered by the water of the dam. The combined flow from the gates exceeds 4500 m³ (one million gallons or 4,500 tons) of water an hour. Irrigation system is flood-flow at over 8 ha (20 acres) an hour — one man control.

PLATE 11

The same type of watergates in an irrigation channel release water into an irrigation bay. The earth banks at left foreground and right centre are parts of the adjacent steering banks which form the bay, foreground, being irrigated.

from a watercourse. It may be practical to lead water from a watercourse along a water race to a rise close to and above the site, and from that point flow the water over the wall through fluming.

Filling is controlled by a low weir wall constructed across the supply stream bed. A suitable notch outlet is provided to control the water. The wall, constructed of logs, grouted stone or cement, need be only 0.6 to 0.9m high.

The fluming for the complete ring dam may be made of a variety of materials, but its shape is always that of a long trough. Wood or iron fluming is most suitable and the fluming is supported by a trellis of bush timber.

See sketch and section of contour, broken ring and complete ring dams on pages 64 and 65.

The dams, ranging from the high contour to the complete ring dam are suitable for easy construction and very profitable use in a wide variety of farming land. Small bulldozers may be used. All the land that can be spray irrigated from such dams will develop rapidly in fertility, productiveness and value. Keyline progressive soil development, greatly stimulated by the correct use of spray irrigation, will bring such land very close to the value class of fertile irrigable river flats.

The overall costs of spray irrigation will be less than those pertaining to river flats and the pumping of water from the river. River water will have to be "lifted", whereas the water of these dams is at least "assisted" by gravity.

CHAPTER NINE

Trees

TREES on land originally timbered were part of the natural soil development. In no circumstances is the complete destruction of all such timber necessary or desirable for farming and grazing.

There is probably no other land development work which has been so completely unplanned and haphazard as that of timber killing and clearing, and no factor of fertility so completely ignored.

To grow crops and satisfactory pasture on forested country, clearing of timber is necessary. Gradually more and more timber is cleared because of the disadvantageous effect of trees on crop land.

However, like cultivation, clearing has been overdone, with the result that soil fertility eventually suffered and crop and pasture yields were affected.

Grasses and timber do not usually grow well together. A large tree will all too often affect quite a sizable area of crop or pasture land and the tendency is to get rid of the tree.

On some farming lands trees are left scattered about. No longer living in forest conditions, they tend to die. It is often observed that the upper and outer branches are dead; the trees are slowly dying together. On some farms they are already dead.

Properties which contain some steep country are often cleared to allow all the flatter country to be cropped. The steep land is left timbered and used as grazing areas.

The general practice of leaving all steep country in timber to protect it from erosion has not been successful, nor has such practice improved the timber. Steep country, left fully timbered, is often the greatest bushfire hazard and the worst area for pests. A fire in a timbered area, followed by heavy rain, is one of the causes of widespread land erosion.

To derive the greatest benefit from timber for soil fertility and better farm working and living conditions, trees must be left to serve the whole of the property.

Properly located trees cool a property for stock in summer and warm it in winter. They protect the land from winds and in their widest aspect may be capable of some overall improvement in climate.

Keyline timber clearing is planned to derive the greatest benefit from trees for the whole of the farm.

First, trees are left in strips or belts wide enough to keep some semblance of forest conditions in the timber for its normal healthy growth.

Steep country is not left in full timber, but partially cleared and timber strips are left to serve as wind protection for the property.

The Keyline is again the planning guide for clearing. The first timber strip, 10 to 20m wide is left just below the Keyline and forms a Keyline Timber Strip.

In most areas the lower side of that timber strip is suitable for a farm road, being drier generally than the land above the timber strip. Crop or pasture suffers more from the effects of moisture lost to the trees on the lower side of a timber strip. However, when a road follows along the lower side of the strip the little extra water runoff from the road causes both grass and crop to grow well right up to the road.

The timber strip or the road along the timber strip forms a permanent guide for Keyline cultivation.

From the Keyline both up the slope and down the slope of the land, timber strips are left (or planted) on the contour at regular vertical intervals apart. The important guide for determining the vertical interval between timber strips is related to the height of the trees. If trees are about 13m high the timber strips could be roughly 12m apart vertically. That provides some overall wind protection for all the land and locates the timber strips closer together in the steep country and farther apart as the country flattens.

Even in very flat country of low scrub or mallee only 3m to 4.5m high that formula for clearing will provide greatly improved farm conditions.

The only trees necessary other than those on that pattern are the ones left around the boundary of a Keyline paddock area.

Timber strips left as described are a valuable aid to soil fertility, apart from the supply of the deep minerals which they bring to the surface. In wet weather cattle will stay on soft pasture ground only long enough to feed and then return to the firmer ground in the undisturbed soil of the timber belt.

The two most efficient land compacting implements are the sheepfoot roller and the multiple pneumatic wheel roller. The farmer has to contend with his own efficient compactors, which are his stock and wheeled farm implements. The comfortable conditions of the timber strips will keep his stock off wet, soft ground to a large extent. The farmer, of course, should leave his wheel machinery in the machine shed when the land is wet. Thus compaction of the soil, one of the great destroyers of soil fertility, is minimised.

By clearing the steep country on that pattern, more and better grass areas are available and better timber will grow in the timber strips.

Very short steep slope country is always of greater value when cleared and Keyline developed. Suitable timber strips are left on the flatter top country above.

Keyline absorption-fertility methods above the timber strips do, by the greatly increased moisture-holding capacity of the land, provide the timber with better moisture. Timber growth is considerably accelerated.

Timber strips will prevent land slips on country which would tend normally to slip when fully cleared and saturated in heavy rains. The timber strip is a definite and effective anchor, holding the land together.

Keyline cleared land, when subsequently subdivided into paddocks, will have some shelter timber in all paddocks. Every paddock, whether in the steeper slopes or the flat country, can be rotated to grasses and crops in turn.

The only way to ensure perpetual timber is by providing conditions to allow trees of all ages to grow together.

If each paddock in turn is closed to stock and cropped for two years or more in each 10 or 12 years, young trees develop in the timber strips and permanency of timber belts is secured.

To sum up the simple plan of Keyline timber clearing:

Decide on the location of the largest paddock area — see other comment in the chapter headed The Plan — and locate the Keyline or Common Keyline of that section. Then peg or suitably mark a strip or belt from 9m to 18m wide parallel to the Keyline below it. That belt is to remain in timber.

Next mark out the first timber strip above the Keyline by pegging or marking a contour line at a vertical height above the Keyline about 10 percent less than the height of the Keyline trees. Mark another contour line above that one 9m to 18m wide. That pegged area is the timber strip which is to be left there.

Continue that contour marking, both above and below the Keyline.

The contour marking of the tree strips leaves the strips themselves of uneven width.

If tree strips of even width are desirable, then a contour line forms the lower line of the strips above the Keyline. A line, parallel to it, forms the upper line. Below the Keyline the upper line of the strip is on a contour and the lower line is parallel to it.

A strip of trees may also be left around the boundary of the area.

When the country is cleared on that pattern, the timber strips form permanent markers for all Keyline cultivation.

No land could be more spectacularly beautiful than the timbered undulating country of Australia which has been cleared and developed by Keyline planning.

However, large areas of land which will come up for Keyline development have had too much of their timber removed without plan, and growing timber strips will be a necessary part of the best Keyline development.

Generally a small Australian native tree will cost 65c each in batches of 20, or 90c each to plant, but may cost many times that to maintain for a year. While the cost of planting is not so serious and can be reduced by growing the young trees on the farm, the cost of growing timber strips of thousands of trees is impracticable unless some cheaper and easier methods are devised.

Keyline planning and development will permit closing paddocks from stock for two or three years while crops are grown. That time will allow a planted or "induced" timber belt to develop to a stage where the trees will survive without attention.

In large or small paddocks without trees that are to be Keyline conversion-cultivated a timber strip four to 10 tree rows wide can be planned. After the paddock has been completely cultivated tree rows are marked; the first row by a deep single rip cultivation parallel to the Keyline or Guideline. The distances apart of the farther rows of trees are gauged by the tractor which will later cultivate between the rows. The following procedure has been found suitable.

After completing the full Keyline conversion cultivation of the paddock, mark out by a single rip the first tree row position. A single shank is allowed to penetrate deeply through the ploughed soil. On the return run with the tractor, place the higher side rear wheel in the lower wheel track of the first run and travel the tractor back without ripping. Turn and again with the uphill side rear wheel in the lower track of the last run, mark out, by ripping deeply, the second tree row. Repeat to the number of tree rows to be planted. That row spacing will allow the tractor later to cultivate satisfactorily between the tree rows. One or two cultivations are advisable during the first year.

The work is done some months before the time for planting the young trees, so as to collect as much deep moisture into the earth as possible. The object is to improve the soil and to provide sufficient moisture in the soil before planting the young trees, so as to avoid entirely the necessity for watering later.

Australian native trees should be planted when around 50-180mm high and a few months old, and planted directly from the tubes as used by the forestry nurseries. Plant the young trees well into the moisture zone without breaking the tubed soil in which the tree was grown. Press the soil down very firmly around the trees.

Trees can be planted quickly into the deep moist soil with very few losses and without the addition of water. The distance apart of the trees in the row may be closer than is intended for the developed trees. Spacings of 2.4m are suitable for a variety of tree species. Planting time varies in different districts.

If watering and hand cultivation can be avoided, the chief cost of growing the trees is also avoided.

A tree strip on a Keyline may sometimes be satisfactorily grown by planting the tree seeds directly into the paddock.

Trees can be induced to grow by a variety of means without actually planting young trees or tree seeds, by merely leaving a strip of country out of ploughing when the paddock is closed for cropping. Tree growth will often flourish on the untouched strip and form a valuable tree strip.

Two interesting incidents are recorded here to show that other low cost means of growing valuable timber strips are available to the farmer:

1. During the construction some years ago of several water races, the completed drains, all except one, were harrowed and fertilised. A directive was given that one drain was not to be treated or touched in any way, to see just what would grow on it. A variety of rubbish grew quickly on the exposed subsoil. Three years later a row of trees 6m high, all of one species, covered the drain.

2. During a very dry period several runs with a heavy road plough were made to form a fire break. Later the dry grass of the fire break strip was burned off. The paddock was not stocked heavily during the following two or three years. At the end of that time the fire break strip alone was then well overgrown with trees all of one species. The trees here were a different species entirely from those growing in the drain about a kilometre away.

From those happenings it can be seen that whenever a treeless paddock is to be closed up for cropping for two years or more, a suitably marked and planned strip of land should be left untouched, or perhaps given some special attention so as to allow a timber strip to develop of its own accord. Once the trees are two or three years old most will survive stock damage.

CHAPTER TEN

Steep country and valleys

CLIMATIC FEATURES have a profound influence on soil development. Gently falling rains are better for natural absorption-fertility than sudden heavy downpours. The gentle rains are absorbed into the ground with all their fertility factors. Ground moisture lasts longer and beneficent decay continues while moisture is present. The surplus water percolates underground after the majority of fertility factors are filtered from the rainfall. All the gases contained in the rain become available in sufficient or maximum quantities to aid optimum fertility development. Many kinds of basic minerals, organic elements and chemicals are contained in the air.

There is every reason to assume that a wide variety of elements are brought into the soil when rain is absorbed and held long enough to enable the humus of the soil to filter them into itself.

In the harsher climatic conditions generally affecting most of the Australian farming lands, natural absorption-fertility does not apply to the extent it does in countries of milder climatic conditions. Moisture losses continuously retard decay. Methods of extraction-fertility farming act more rapidly to reduce soil fertility to the stage of active soil erosion. Australia, of course, is not alone in that America and Africa have similar conditions. The causes of erosion are precisely the same in those and many other countries.

Just as obviously absorption-fertility farming on those lands will be more strikingly effective than in the countries of benevolent climatic conditions. If, by every practical means, rainfall is absorbed into the earth and all its fertility elements held, and if evaporation losses of moisture are retarded greatly and processes of decay continue longer, then countries with harsher climates may develop more rapidly in fertility than the others. In such development, the most important type of country of all is unquestionably the steeper mountain and hill lands.

The effective control and rapid soil development of those lands will do much to mitigate the calamitous effects of the worst feature of our climate. The worst effect of droughts and flood can be fought and conquered by absorption-fertility methods of farming and grazing if applied quickly to steeper lands.

At the present time the rapid runoff from those lands directly causes uncontrollable and destructive floods, with losses of life, alarming destruction to property and stock, and the choking of rivers and harbors with silt. The trail of havoc extends from the mountains to the sea.

In dry, hot weather the steep undeveloped and uncontrolled lands are a constant menace with their bushfire hazard. Extermination of rabbits and other pests is more difficult in these lands.

Present recommended orthodox practice is to fence well, stock conservatively and leave the timber on these steep lands to protect them from erosion.

The trees of the steep grazing lands of Australia do not of themselves prevent erosion. Only good management does that. Timbered areas require better management to protect them than do grasslands.

Generally, the present condition of steeper country is such that it will not grow both good grass and good timber.

A Keyline principle is that planning and development above the Keyline comes first. Development must be sufficient at least to protect the lower lands.

The full development of steep country can only be accomplished rapidly and effectively if it can be made to pay. The profit must be almost immediate and definite — not something in the dim and distant future.

The first approach is simply to enable the steep country to absorb the rain that falls on it and keep it in that condition. Then follow the Keyline method of soil improvement for pasture lands.

Clearing of that country on the Keyline clearing plan leaves permanent timber strips which form a guide for Keyline absorption-fertility cultivation.

The full control that gives will enable the effective control of bushfires. The parts of the steep country that may be adjacent to an outside fire risk area can be managed to protect the whole property. It can be heavily grazed or cultivated to form a completely effective fire break.

Considerable areas of such steep country, often regarded as of lesser or almost insignificant value, will be found to develop better pastures than the land now considered as the best on the farm.

At present, when a farmer leaves his steep country in timber it is usually because he feels he must do so. Sometimes it is left because he really wants to run it as a forest for profit or for general farm purposes. He thinks then that the steep country is the only place for such a forest. In the Keyline development of steep country the farmer has the choice. He can develop high quality soil and pasture, or if he wants a forest area he can have it in the steep country or *anywhere else.*

Australia is, geologically, a very old continent. There are no very high mountains and practically no steep country of long, unbroken slope. By steep slope country is meant slopes of 100 percent, or a rise or fall of 0.3m for each horizontal 0.3m. Many slopes of 100 percent do exist, but they are nearly all short slopes rarely 45m long horizontally. Such short steep slopes generally exist as the sudden dipovers that form the valley heads — the start of the valleys. If a farmer wants some good growing timber he will rarely grow it on that short steep country. Much better timber will grow on the flatter country above and below.

The clearing of timber on sloping country is dealt with in Keyline Timber Clearing. Slopes of up to 0.3m rise in 0.9m are Keyline developed as described in earlier chapters. Most wheeled tractors will do that safely and well. The three-point hitch and other tractors on which implements are mounted, especially those of about 30 horsepower, are particularly effective.

For short slopes steeper than one in three and up to one in one, a crawler tractor and a trailed Yeomans will provide means for full development of soil for pasture growth.

Keyline cultivation of the type required, up to three workings, is first completed above the slope to 9m from the steep dipover. The slope is then ploughed directly down hill. The tractor and plough make the turn in the flatter area below and then travel squarely up the steep slope in the same width of ploughing of the downhill work. Turn in the flatter area above and plough directly downhill beside the ploughed ground of the first downhill ploughing width. Continue ploughing directly down hill in new ground and uphill in the ploughed land of the previous downhill width. By "breaking ground" only on the downhill work and travelling uphill in the ploughed ground, the load on the tractor of the uphill work is reduced to such an extent that the tractor will handle the implement on the uphill travel without continuous implement depth adjustment. The whole of the steeper country receives two cultivations by that means, and that completes the steep country ploughing.

The next step is to start from the Keyline and plough the area above the Keyline to the bottom of the steep country ploughed area. Follow Keyline cultivation parallel to the Keyline on the last run.

Such land may in a short time grow some of the best pasture on the farm.

Unless it is solid rock, there is probably no country too steep for improvement if stock will climb it for food. Country which carries some soil, even if it is very steep, will display significant improvement by Keyline full development undertaken in the area immediately above it.

It will probably be much richer in the minerals of fertility than land which has been heavily cropped and grazed by methods that have not only been extracting fertility but destroying more fertility than they extract in crop and stock production.

Our droughts and flooding rains are factors of prime importance in the loss of fertility and later in soil losses by erosion.

The effects of both are capable of profit producing control, by the absorption of fertility into the soil of the hill lands.

Valleys start or head where a portion of a slope near the top of a watershed or divide becomes steeper than the general slope on either side. Thus the first part of a valley formation is steeper than the ridges or shoulders on each side that form the valley. At some point down the valley — the Keypoint — the valley slope flattens to such an extent that it becomes flatter than the ridges or shoulders on each side in the same vertical interval.

Those factors, as already stated, locate our Keylines. The valley itself and the ridges which form the valley are the two points requiring special care once the steeper land above has been controlled by Keyline absorption-fertility. The valley floor has been a danger point for erosion and may have gullies which require repair. The shoulders of the valley are usually the driest areas.

While the valleys of the usual farming and grazing property, if not eroded, carry more moisture than other areas, they continue to extract ridge moisture even when the valleys are too wet for healthy growth. This "dog in the manger" aspect of the valleys is quickly offset by Keyline cultivation. The ridge areas then receive and retain their fair share of moisture for longer and longer periods.

If the valley is eroded the erosion holes will continue to bleed moisture to the atmosphere until little remains. The effect of that process can be observed where every tree of a forest is dying in an eroded valley area.

Sufficient has been said in earlier chapters to illustrate the effectiveness of Keyline cultivation in the control and development of absorption-fertility in valleys. Where significant gully erosion exists it can also be controlled by correct Keyline methods.

Keyline development first controls the usual water runoff into valleys from the higher land, by tremendously increasing the absorption capacity of the area and diffusing the excess water, thus greatly retarding and delaying its concent-tration time.

With the high country in that condition effective gully control and repair in the valley below is simplified. If the hole or gully is not large, repair is best done without the use of bulldozers. Repairs with those implements, where the valley soil is not deep, too often result in the topsoil finishing in the hole and a relatively large area of slow-to-improve subsoil remains. That will spoil the look and value of the repaired valley.

A much better procedure is to use the Yeomans plough for the repair work.

Plough up each side of the valley, allowing one end of the "plough" to drop over into the hole. Spikes with heel sweeps quickly move sufficient soil and subsoil from the banks into the hole and allow the "plough" to cross in all directions. Only sufficient filling or levelling of the hole is necessary to ensure that the deeper part of the hole is not lower than the valley immediately below. In that way ponding is prevented.

The repaired valley does not then expose all subsoil. The soil and subsoil will be mixed and the effect of a bare infertile patch in the valley will more quickly disappear.

Immediately the repair is complete the area is Keyline cultivated or pasture improved and the repaired area is practically safe.

It is seen that the procedure is as follows — assuming the holes are below the Keyline. First, complete the cultivation above the Keyline. Second, plough in the holes. Third, immediately Keyline-cultivate the area from the Keyline parallelling the Keyline downward.

If the area treated is poor heavy soil, very low in humus content, it will tend to seal quickly from very heavy rains. When that happens the area is treated again both above and below the Keyline on the procedure already discussed for soil improvement of pasture lands. That is necessary to provide oxygen so improvement will continue. Without oxygen both decay and growth will tend to cease. As the soil improves less cultivation will be necessary to provide aeration and absorption. The improved soil will provide those itself.

The great advantage of Keyline aeration cultivation on soil that is sealed is sufficient to warrant its use even if a recently sown pasture is still new and weak. In those circumstances the cultivation still follows the Keyline procedure. Spikes are to be used and spaced 610mm apart. One fast run completes the necessary Keyline aeration cultivation and soil improvement will continue without check.

If the erosion holes and gullies in a valley to be Keyline-improved are merely to be "killed" or prevented from getting worse, the procedure is the same,

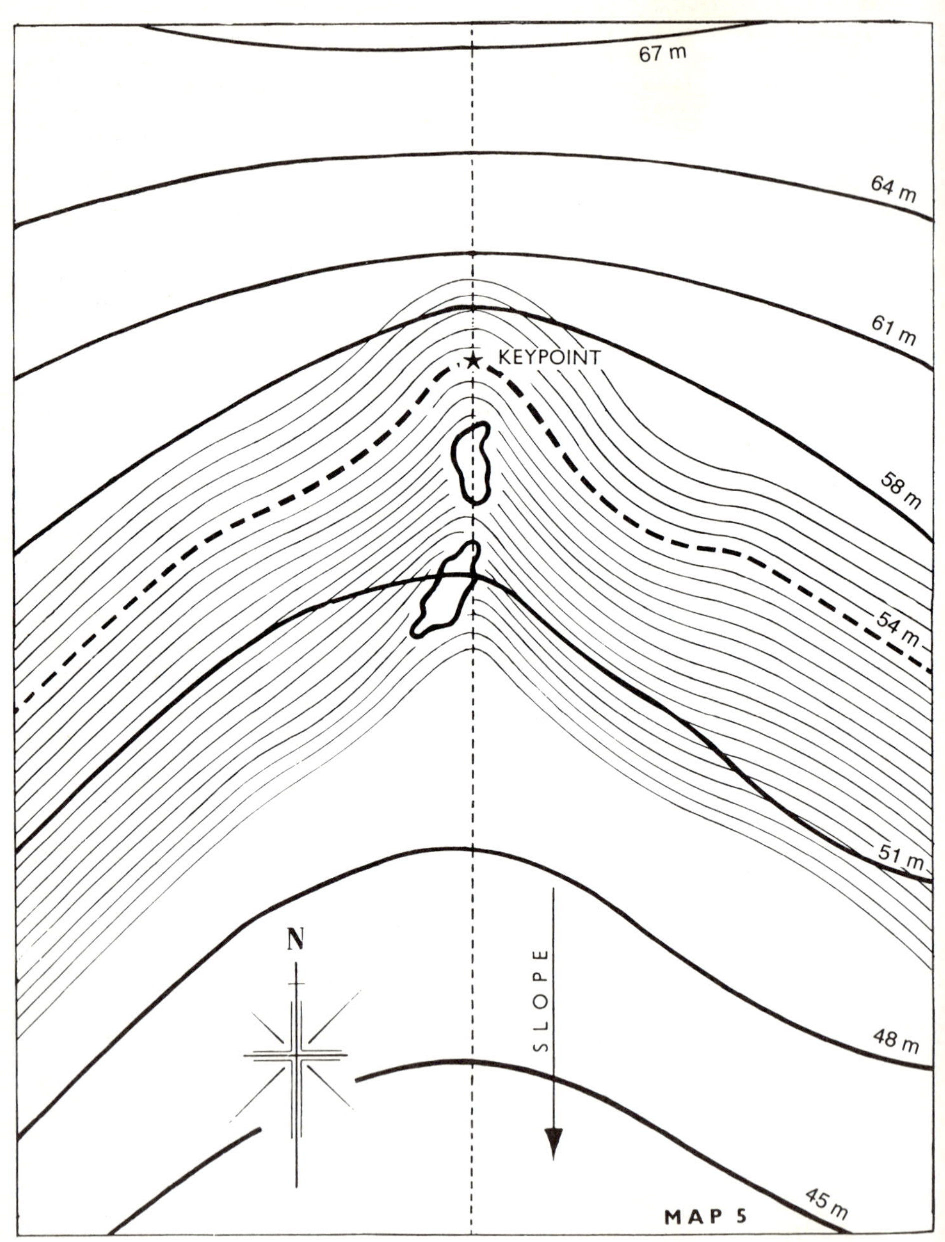

67 m
64 m
61 m
58 m
54 m
51 m
48 m
45 m
KEYPOINT
N
SLOPE
MAP 5

except that the Keyline-cultivation — that is the final cultivation run — is divided or split to suit the circumstances. See Map 5.

Parallel the Keyline progressively, crossing and re-crossing the valley until the first erosion gully is encountered. Then complete the parallel runs downward, working to the gully and back on one side until that side cultivation reaches the end of the hole. Continue the cultivation on the other side of the hole to the bottom — downstream end — of the hole. The next cultivation run will then be a complete one parallel to the others and again crossing the valley, but now below the hole.

All the cultivation running from the erosion hole out and away from it will have the Keyline drift away from the hole. Only with very heavy rain will water again run in the gully, and if any more erosion takes place it will be on a greatly reduced scale. With continued Keyline soil improvement it will cease altogether.

In times of severe drought the only noticeable green tinge on a grazing property will often be the narrow moist strip in a valley. One of the first very noticeable effects of Keyline development, if followed by severe drought conditions, is the greatly widened area of longer-lasting valley greenness.

Keyline soon multiplies the effects of the average rainfall.

Practically no valleys are safe from erosion under "extraction-fertility" methods of farming and grazing, while under methods of Keyline absorption-fertility all valleys, including those of the steeper land, are safe from soil losses and consequently erosion. Not only are they safe, which is a negative matter, but they will improve progressively with all other land in fertility, productiveness and beauty.

The use of trees in Keyline planning is discussed elsewhere in this book, but a special reference to trees and their effect on valleys can be considered here.

In the Keyline development of land, trees are not generally left in valleys except as part of a Keyline or guideline timber strip. The eddying of water caused by a tree in the path of the water flowing down a valley will often start an erosion gully. The breaking of the soil round a tree from root movement and growth can also be a contributing factor.

Stock sheltering beneath a tree tends to powder the soil around it, thus causing soil movement when water flows down the valley.

The effect of both erosion hazards will be quickly offset by Keyline improvement. It is still advisable to clear the valley timber except when a Keyline or guideline timber strip crosses the valley. Stock will not damage timber left in a valley as part of a timber strip crossing the valley.

The Keyline development of valleys is simpler and more rapidly effective if there are no odd trees to consider.

It has been noted that a mob of cattle in a large paddock containing three timber strips at different levels invariably all camp in the one timber strip and spread themselves well along the line of this belt. A night or two later they will be together maybe in a higher or lower timber strip.

CHAPTER ELEVEN

Flatter lands

IT HAS BEEN PREVIOUSLY stated that Keyline cultivation parallels from the Keyline up the slope of the land and from the Keyline down the slope of the land. However, there are very many properties which do not contain Keylines or a single Keyline, and so a means of the simple application of Keyline cultivation on such lands needs a guideline on which to work.

Those areas of farms are treated in the same way as are all areas below the Keyline. Cultivate the land parallel to the highest suitable guideline, always working parallel down the slope of the land.

The line that forms the overall or planning guide on such properties is called a general guideline, and, as with the Keylines, may be either a selected true contour line or a line with a very gentle slope. The slope would be for a water race connected to a water storage.

The special or significant feature of all land lying below the Keylines is that the valley slopes are generally flatter and wider than the adjacent ridge slopes which form the valleys. It was fully explained in Chapter 3. The aim of Keyline cultivation is the equalising of the moisture between the wettest and the driest parts, that is between the valley and the adjacent ridges. To do so most effectively a guideline is located in the highest position, where it can serve as a guide for Keyline cultivation.

If the slope is long, another guideline at a lower level is located. It lies at a convenient distance below to serve as a boundary to the upper area. It is a lower guideline, and is usually a true contour line. It is marked by any suitable means, preferably one permanently locating it.

The control and development of those areas is approached first from a consideration of water which flows down to the valleys from the higher country outside. The entry of that runoff water is usually at the lowest point along the highest boundary fence. That may also locate the guidepoint from which a level or sloping line in both directions suitably forms the general guideline.

If a large area of land lies above the selected general guideline it will be necessary to locate an upper guideline to control the Keyline improvement of the higher area. If so, the upper guideline is located and marked as high in the area above the general guideline as possible. Care should be taken to see that it is of sufficient length to serve its purpose.

Outside runoff water may now be a problem. Perhaps the main factor in determining the general guideline will be the position of a suitable conservation dam site for the storage of that extremely valuable water. The site is looked for in the highest third of the area, and when located the general guideline becomes a suitable water race to the dam site.

All details of farm planning above the Keyline also apply above the general guideline of the land below the Keyline.

The main grazing or large cultivation area is below the general guideline. A lower guideline located at a suitable distance below forms the top boundary of another group of smaller paddocks. If their vertical distance below the conserved water is sufficient, gravity spray irrigation is always planned. Five percent of a grazing property suitably planned and supplied with water for gravity spray irrigation may add 50 percent or much more to the capital value of the whole property.

In the development of timbered areas of that type of country, clearing is done to leave suitable timber strips along the general guideline and all guidelines.

The formula mentioned in Chapter 9, which relates the vertical distance apart of the tree strips to the general height of the trees, is again the planning guide.

Map 6 illustrates in simple form a valley area below the Keyline and the location of the guidepoint and general guideline. The parallel lines on the map which start from the general guideline and parallel it downward illustrate the drift of water out of the valley. It compensates the natural water concentration in the valley. Keyline cultivation is again completely effective. (See page 86.)

In selecting the guidepoint — in place of the Keypoint of properties containing their own Keyline — it may be advisable to locate it just away from the fence at the lowest point along the highest boundary. A distance of 6m from the fence would allow a farm road to cross the paddock above the general guideline.

Soil erosion by water is simply and profitably cured on flatter lands by the methods of this book.

There is, however, a type of erosion which appears to defy man's efforts to cure it when those efforts are confined to "maximum soil improvement". It is the serious periodic erosion by wind, which occurs alike on poor soils and fertile soils of our marginal lands.

Following three or four years of much drier than usual conditions on such country when it has a normally sparse rainfall, the serious wind erosion manifests itself. If the latter end of a dry period coincides with that of a severe drought, followed by high wind, the soils will move in vast quantities.

The dry period or the severe drought cannot be controlled. The only possible solution to the problem lies in measures designed to retard the ground velocity of the winds. A rough cloddy surface will reduce a 96km/h wind to a velocity that will not raise any appreciable dust from the soil, but at the end of such a period of those conditions the surface condition alone will not have sufficient effect.

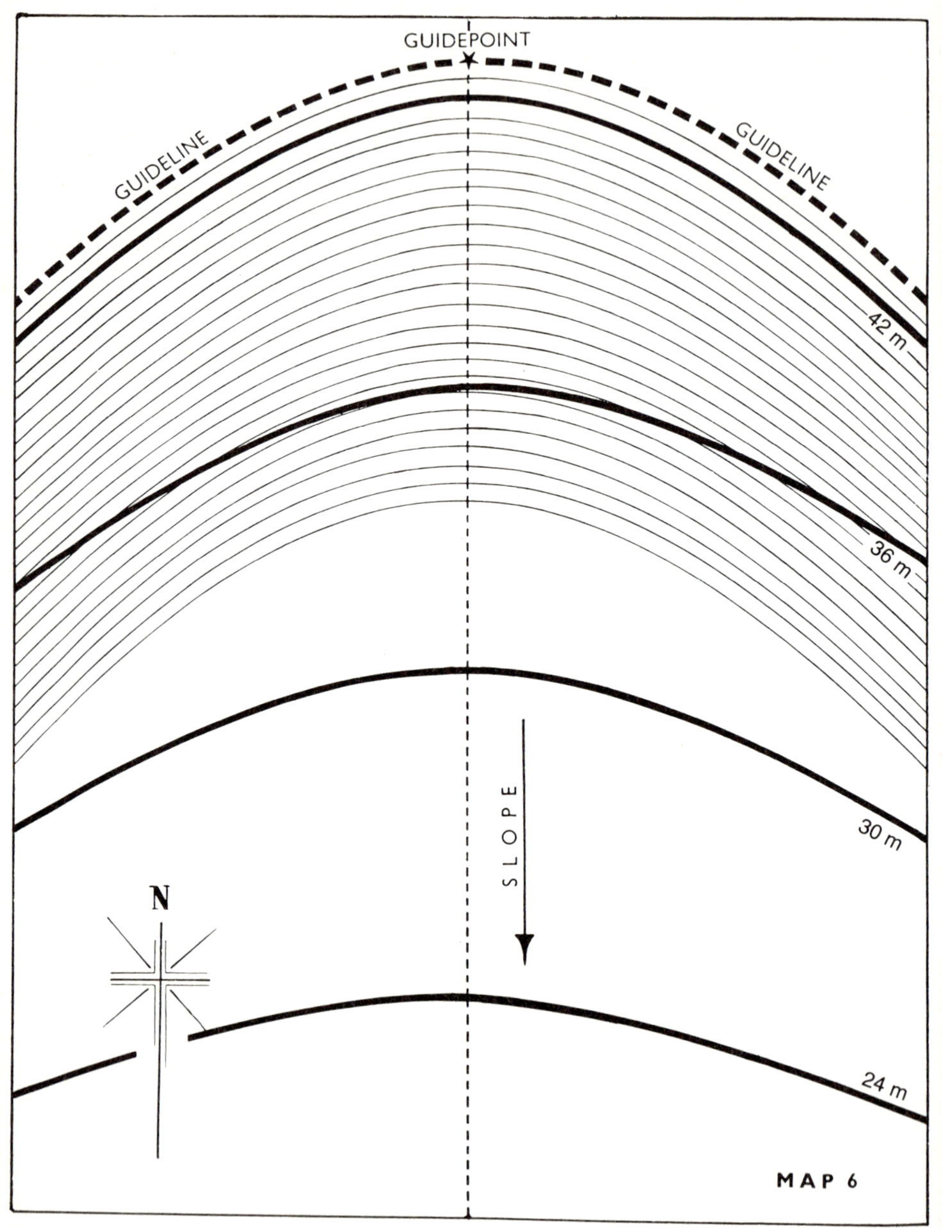

GUIDEPOINT
GUIDELINE
GUIDELINE
42 m
36 m
30 m
24 m
SLOPE
N
MAP 6

Growing sufficient tree strips is the only possible means of reducing the high velocity of winds to such an extent that the soil will not blow. The problem is one of great magnitude and the solution in the planting of trees must be of like proportions.

Indigenous trees can be induced to grow by leaving protected strips of land in the right pattern. That is the lowest cost means of growing the tree strips on a large scale. If the country is treeless, then tree species will have to be introduced which will not only grow well in this country but also survive the period of very dry conditions.

Nothing can be done during the time of the actual blows that will give results commensurate with the money expended. Planning the work can be satisfactorily done at that time so when better rainfall conditions follow the drought the land will be in a position to make quick rejuvenation. Four years later such land could be safe from wind erosion.

CHAPTER TWELVE

Other applications

KEYLINE DISCUSSIONS SO FAR have been concerned with land areas containing valleys. The prime purpose of the lines of cultivation on the Keyline principle is to counteract the natural rapid concentration of rainfall into valleys by an induced drift out of the valleys. At the same time the particular type of cultivation discussed in Chapter 4 enormously increases the absorption of rainfall into the soil, and effectively uses the rainfall for progressive soil development.

It is a practical impossiblity to plough accurately on the contour unless every travel line of the plough is level pegged as a true contour line. That would require hundreds of lines of instrument levelling in every small paddock. When contour cultivation is attempted it must drift mainly off the contour. Contour lines are rarely parallel to each other. They are never parallel in undulating country.

Keyline cultivation, although it may start on the contour, is soon "off the contour" by parallel working. It is the off-the-contour effect which is controlled in Keyline to counteract the natural concentration of rain water in valleys.

The controlled, and completely effective, drift-off-the contour of Keyline cultivation is as fully applicable to areas of land which contain no valleys or depressions.

A paddock area with an even slope in one general direction is best developed from a guideline located as high in the paddock as possible, and one that still gives a line of sufficient length to serve effectively as a guide for Keyline cultivation. Cultivation parallels that guideline down the slope. The small areas left above the guideline are cut out in any convenient manner. If such an area contained small erosion gutters they would be cured by such cultivation method.

Another paddock without a valley may have one side of the paddock steeper

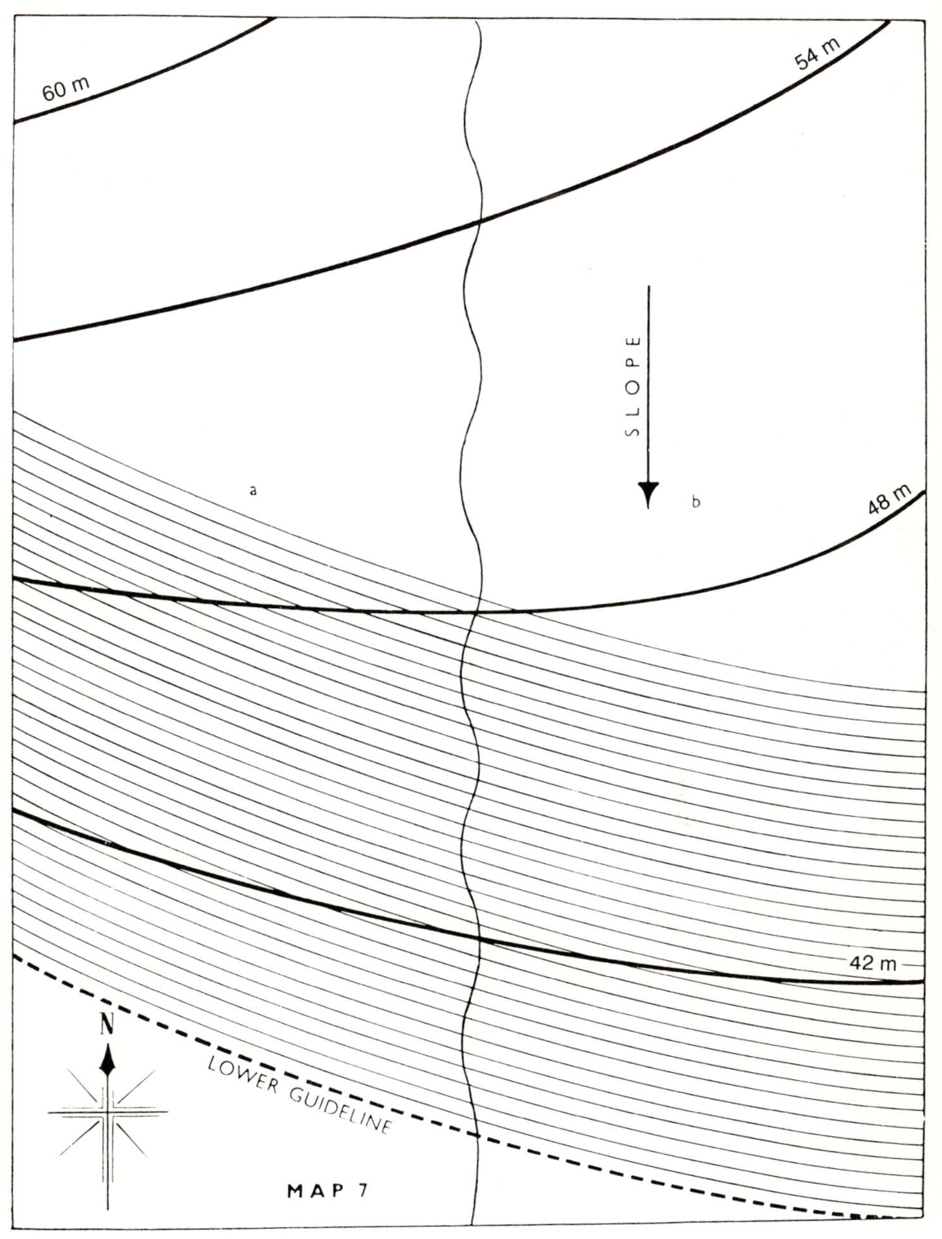

60 m
54 m
48 m
42 m
SLOPE
a
b
N
LOWER GUIDELINE
MAP 7

than the other. It may be necessary to drift the moisture in one direction, while in different circumstances the opposite may be advisable. It can be done by Keyline cultivation, as illustrated in Maps 7 and 8.

Map 7 illustrates a paddock area containing a steeper side, "a" and a flatter side "b". Assume area "a" is partly eroded and the whole paddock is to be Keyline improved. It will be necessary to counteract the fast runoff to the south west from the area by an opposing drift in cultivation furrows away from that direction. A guideline is located, the lowest suitable in this instance, and Keyline cultivation parallels that line up the slope of the land, as illustrated by the parallel lines of Map 7. They have a drift away from the natural runoff direction. Protection and development are thus secured. (See page 89.)

We can now assume an opposite problem on the same area. Area "b" in this case is wet or swampy and "a" is very dry. A drift towards "a" of the surplus moisture of "b" is desired.

A guideline is located in the highest suitable position and Keyline cultivation parallels the guideline down the slope of the land, as in Map 8. The surplus moisture of "b" now has a "drift" to the dry area "a", with the effect that both areas are immediately improved. The small areas left out of parallel cultivation are worked out in any convenient manner. They will not affect the effectiveness of the work.

The sour wet area "b" is properly aerated for rapid improvement and surplus moisture drifts to the area "a" to improve it. Surplus moisture in those circumstances may drift along the tine furrows undergound away from "b" until the area is left nicely moist, as distinct from wet. Moist soil — not wet soil — produces healthy pasture growth.

With an appreciation of the astounding effectiveness of Keyline cultivation and some experience of its use, it will be found that relative moisture content of problem land can be adjusted at will by the astute use of Keyline's off-the-contour type of cultivation.

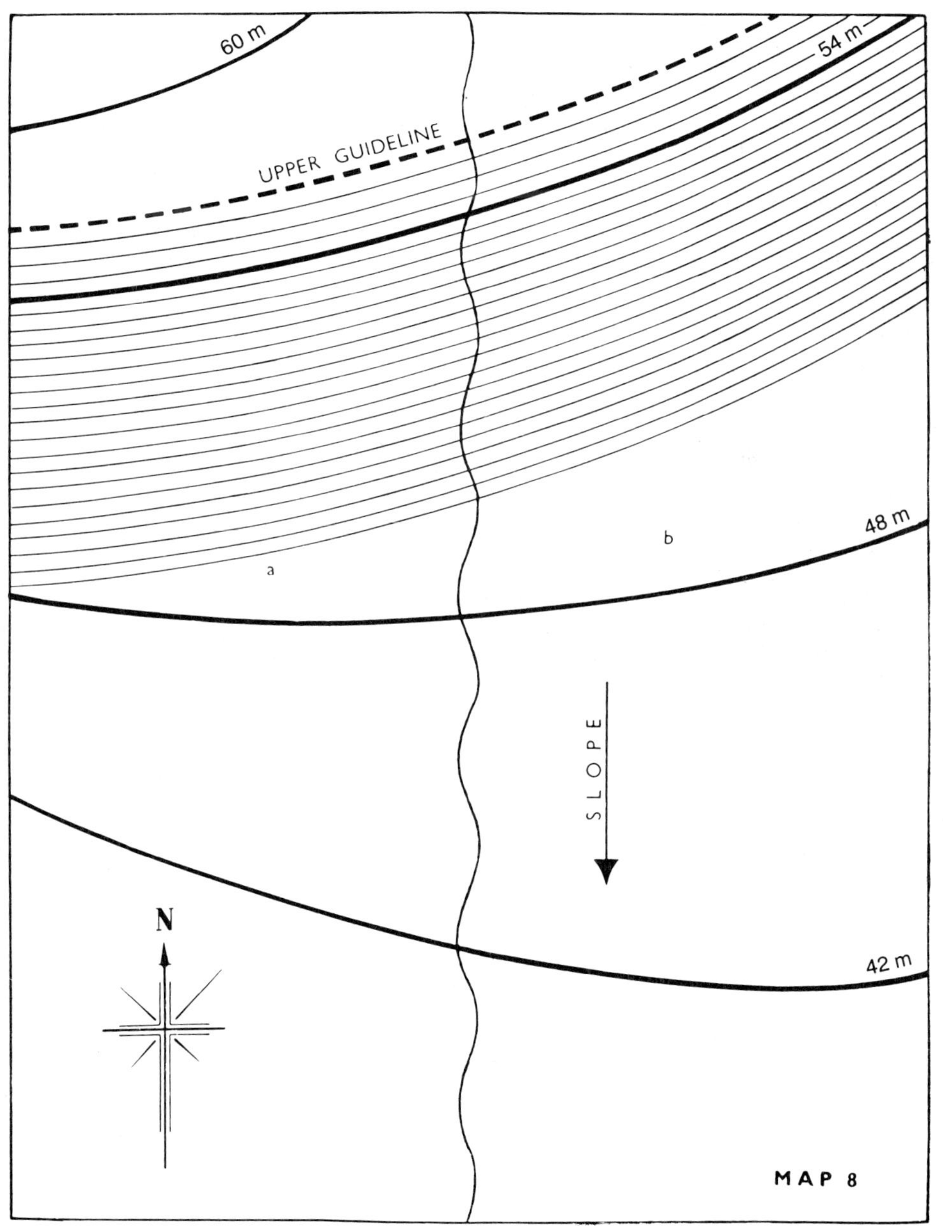

60 m
54 m
UPPER GUIDELINE
48 m
a
b
SLOPE
42 m
N
MAP 8

CHAPTER THIRTEEN

The plan

FULL KEYLINE PLANNING, in the development of farming and grazing land, is the logical use of all the methods of Keyline which have been discussed in this book.

Keyline timber clearing cannot be applied on cleared land, but the design of Keyline clearing to "leave" timber as strips or belts can be applied in growing timber to aid soil development and for general usefulness. Growing trees in suitable numbers cannot be attempted at once over all the farm area, but a tree belt can be grown in two or three years in a paddock conveniently closed for cropping. The immense satisfaction from a successfully grown timber strip in the first paddock would certainly induce the farmer to continue the program into other paddocks.when convenient.

Water conservation in Keyline and high contour dams obviously can be employed only on farms of suitable land formations. Such farms embrace huge areas of the most important land from a national point of view. Not only are those steeper lands capable of tremendous and profitable improvement, but by their effect on all the lower lands in their common catchment area exert an influence over many more people than live on them.

While Keyline dams and the high contour dams of the Keyline plan are limited to properties with their own Keylines, the principle of locating some dams high on the farm is almost universally applicable and profitable. The design and the layout of farms should locate as many of the water-shedding areas and buildings as possible above the dams. That would ensure additional water storage. Many of the dams below the Keyline will provide water by gravity pressure for spray irrigation and stock watering systems.

It is a principle of the Keyline plan that all land on the farm is made to absorb all — or nearly all — the rain which falls on it. Surplus rainfall runs off slowly along the natural flow lines of the land. Water is transferred for storage only and never to another valley for disposal. Rapid runoff and fertile, absorbent soils. In many places damage from present water runoff is accelerating. The Keyline plan first retards and then completely prevents the usual erosion of farming and grazing lands.

Keyline progressive soil development or any other Keyline work, by being complete and fully effective in each area on which it is applied, whether on the small paddocks of a farm or on a large grazing area. requires no outside co-operation or co-ordination.

It is completely effective as an isolated unit.

The Keyline plan used on farms in an area of regional planning is complete in itself. Every farmer, by improving his land, is doing the best possible for the region, but he is still an individual working for his own pleasure and profit.

General land development is always vitally concerned with water. Whether the object is conserving water for the production of soil fertility and increased yields, or whether the aim is the control of water for flood prevention or irrigation schemes, the general subdivision of land into smaller areas and paddocks is best governed by natural watersheds.

Keyline planning of a large area of land first aims to divide the area into smaller units or paddocks which are suitable for later economical development and farm working.

A good contour map of the area is of great value in this planning. A map with contour lines at 6m vertical intervals is suitable for land containing slopes from gently to steeply undulating. Three-metre contours are suitable for gently undulating areas and 1.5m for flatter slopes. On the flatter country contour intervals should be such that at least three contour lines are contained in the large paddock areas. With fewer than three contours such maps do not display a complete picture of the land for subdivision and development. Watershed areas both small and large can be located at a glance. Keylines and Common Keylines are readily found on the map; in fact, the geometry of the contour lines emphasises the Keylines. The steeper country appears to be narrower proportionately between the contour lines on the map than does the country of lesser slopes between its lines.

Maps enable the planning lines to be located in the approximate position in which they will be used in Keyline development on the land itself. Keyline areas, Chapter 7, located from maps, can be readily plotted on the land.

Good farm contour maps as described are, however, rarely available now, but the importance of "planning the work then working the plan" in all matters relating to land development is such that *the use of good farm contour maps should become general practice.* It would be of tremendous benefit to the farmer if some service were available to produce farm maps quickly and cheaply. Parish maps are generally the only ones now available and they, increased to a larger scale, can serve as a basis for mapping the areas. Keylines as located on the property can be plotted on the parish map and so form a simple and effective farm map.

The largest suitable land unit for planned development is that contained in the watershed of a river system. Within that large area of land are contained the many smaller watersheds of the creeks and streams which flow to this river. Again, within those smaller watersheds are the lesser watershed areas of all the valleys which flow into the smallest water courses. The lesser valleys are the valleys of the Keylines with which we are directly concerned in Keyline development. Single valley Keylines and Common Keylines form the lesser subdivision of the Keyline areas (Chapter 7).

When large land areas are cut up for sale they are usually subdivided along the lines of existing fences. As the likely fate of all large good land areas is subdivision into smaller farms, the initial subdivision into larger paddocks can be planned for their later development into separate farms of a satisfactory living area. Watershed areas of the large paddock size may be suitable for the purpose. Good subdivision at that time will further enhance the value of the land when it has been developed.

On undeveloped land, which is many times the size of the potential developed living area, one such large paddock can be fenced adequately and Keyline developed to a profitable farm or grazing property.

Within that area the Keylines are first located. Development then follows the pattern of the various aspects of Keyline; timber strips are located; smaller paddocks are determined; buildings and yards, and so on are located above the Keyline; irrigation areas are pegged below the effective water pressure level of the Keyline dams and high contour dams.

The general picture of Keyline planning in undulating country follows a distinctive pattern. The flatter top country above the Keylines contains all the buildings, yards and their roads, as well as the numerous smaller paddocks necessary for the running of all farms or grazing properties. Tree belts are left in the area as described in Chapter 9. Immediately below the Keyline are the large paddocks for grazing and cropping. The lower boundary of the area forms the top boundary of another area of smaller paddocks. They make use of the gravity pressure of the high dams for irrigation. Timber belts are left on the formula suggested for Keyline clearing.

On that plan rapid Keyline development of the first area should pay for the progressive development of a large undeveloped area of land.

The cost of Keyline land development will be lower than the present development of such areas, but the actual cost of clearing may be higher because of the additional cost of the necessary planning which must precede clearing. Extra cost over the usual unplanned clearing may be involved by the necessary supervision.

On land already fenced there is no need to alter the present paddock layout. As Keyline is generally complete and effective in itself in any area small or large on which it is applied, special fencing is not necessary. It may be necessary to dig under a fence in constructing a Keyline water drain to transport water to the Keyline or other dams.

The Keylines, which are the basis of such land planning, have been illustrated throughout this book on simple contour maps. Keylines will usually have to be located without the aid of maps. When the Keylines of Map 4 are to be located on the land illustrated in the map, but without the aid of the map, the Keypoint is located in the first valley. It is done by walking down the steeper head of the valley floor. That is the point at which the valley floor first becomes as flat or flatter than the adjacent ridges.

That point, the Keypoint, is marked by a peg or stake in the centre of the valley. A line of levels, on the longest possible convenient sighting with the levelling instrument available, is then made to the boundary fence in one direction and through the valleys in the opposite direction. When the line of levels reaches the second valley it crosses it on the approximate Keyline of the second valley, and similarly, in the third valley.

At the fourth valley it would be obvious that the line is well below the Keyline of the valley. In this fourth valley a new Keypoint is located and a new Keyline extended to the boundary.

With the line of pegs as a guide, the location for all the Keyline dam sites is studied. If one dam only is to be constructed, the site in the first valley is selected. The reasons for the selection are given in Chapter 8. The working Keyline will then be a drain to carry water to that site. The slightly higher position of the Keyline in the second and third valleys, made necessary by the fall in the Keyline drain from those valleys to the first one, does not present any problem. It can be taken as a usual rule that the Keylines tend to fall in the direction of the general fall of the country.

The actual position of the Keyline drain or other "marker" for the Keyline on the land can always be located or adjusted a little to suit overall circumstances.

The Common Keyline of two valleys may be made to serve the purpose of a common Keyline of three valleys by a little adjustment in its location.

While accurate levels are essential, exact location of the Keyline is not necessary. It is the fact that the aggregate of all the cultivation runs parallel the Keyline and drift down and away from the valley that gives Keyline cultivation its powerful influence.

Referring to the area above the Keyline, Map 4, it will be seen that the land may be developed very rapidly by Keyline absorption-fertility to a state where greatly reduced runoff water is available to fill the Keyline dams below it. Full use of the runoff water from buildings, yards, road, and similar which would be suitably located here, will supply the water to fill the dams. The road alone will shed a large volume of water.

The Keyline plan first develops fertility by maximum absorption in all pasture crop and forest land. The development starts in the steeper areas first. The other great aim of the Keyline plan is the conservation and profitable use of all water that flows to or on the farm. There is, however, no suggestion that large areas of land should be left undeveloped so as to provide a catchment area to shed water for conservation in dams. The use of the water to develop high yields on one portion of the farm at the expense of the larger undeveloped catchment area is completely unsound. That is not the way to either full progressive soil development or maximum yields and profit.

Keyline and high contour dams for water conservation are located in the best possible sites for the effective and low cost application of the conserved water. Gravity pressure for spray irrigation and other purposes is much cheaper than pumped water.

The other dams mentioned in Chapter 8 are placed as indicated. The type of dam to suit the topography is obvious from the discussion in the earlier chapter. The overall aim is again the conservation of all the water that flows to and falls on the property.

First, conserve all the rainfall possible into the soil for the benefit of all the land, and for the production of high fertility. Second, conserve all water which flows from any and all high sources into the highest suitable sites in the Keyline — high contour and guideline dams. Third, provide for other and large storage capacity in lower sites in the contour dams of Keyline, the lower valley dam and the creek or stream dam.

From the economic aspect and the working of a farm some water storage must be provided.

The retention of more water in the soil by correct cultivation methods will provide extra profits. They should be used to pay for the capital cost of suitable dams for irrigation. In turn that will provide more profits.

An overall scheme of maximum water storage can be undertaken on limited finance when each new storage in its turn is used to promote soil improvement and more low cost high yields. Any expenditure incurred in the construction of such a scheme of progressive water storage, including the drains for conserving or conveying water, is deductable in arriving at the taxable income of a primary producer. A farmer in the first year, can claim one tenth of the expenditure incurred, and the remaining nine tenths at one tenth a year over the succeeding nine years. If he ceases to be a farmer, he loses the unclaimed balance of his deductions.

CHAPTER FOURTEEN

Floods or Keyline?

FERTILE SOIL GROWS good grasses and crops, which in turn feed and make healthy animals. The products from those are the dominating factors in the health of the community. Poor soil grows poor grass, poor crops and animals, and they have a detrimental effect on the health of the people.

The vast difference in the flavor of salad vegetables grown on fertile and infertile soil should have been noted by everyone. The products of fertile soil sustain healthful life. The growth from poor soil is suited only to be again absorbed into the soil to help cure the ills of the soil.

The good farmer, by cherishing and improving the fertility of his own soil, is safeguarding the basic factors of the health and prosperity of every section of any community. At the same time he is in the first line of the general fight against disease.

Fertile soil is the basic factor in the health of the community. It is also of the greatest importance to the safety of all the land; it resists to an astounding degree the forces of soil erosion.

There are many other causes of soil erosion than those which may originate from the actions of our few generations of farmers and graziers. While no one generation of farmers caused a significant amount of soil erosion, the accumulation of soil damage from past generations has manifested itself in greater soil movements in the last generation. The forces of erosion are accelerating.

Whenever runoff water is artificially concentrated, an erosion hazard is created. Damage from public roads and other sources completely outside the responsibility of the farmers and graziers cause widespread erosion on the farmers' own lands. Government stock routes and forests are not free from erosion. A bushfire from any cause is always a hazard. A careless camper, a cigarette from a motorist, a piece of glass focussing the sun's rays — all are serious in accelerating soil erosion.

There is, however, no doubt that concerted actions by the community of farmers and graziers could do more in much less time to stop erosion and the shockingly devasting floods, than all the authorities concerned, even with un-limited money.

It would take at least two years for the various authorities who would be concerned to agree on any plan. The work could be completed by the farmers in that time. They would incidentally have increased the value of their land and made additional profit.

To be quite specific, if the Keyline plan were adopted by the farmers and graziers of, say, the Hunter River Valley, the result would be certain and rapid.

Every farmer and grazier would enrich himself greatly by the resulting in-creased value of his land and the better quality of his farm yields. The whole of the Hunter River and its eroding banks and flats would be protected by the farmer's work on his own land. Devastating floods would not occur again at such important population centres as Maitland or any other town on the river. Clear water would flow in the river all the year round and the flow would be more even and constant.

If we assume that the ancient flow of generally clearer water was compatible with the early better anchorages in Newcastle Harbour, may not a new flow of cleaner water result in gradually clearing the harbor, instead of the present continually increasing depositions of silt? Would not a constantly greater flow of cleaner water result in the removal of recently deposited silt from the lower reaches of the river?

All the huge water conservation projects and all the special dams for flood mitigation will not hold as much water as the land itself if all the soil is kept in a condition to absorb the rain when it comes. Dams for flood control are effec-tive if they remain only partly filled, so that large potential storage is always available to act as huge shock absorbers for the floods.

To the *new* vast water storage capacity of the soil we must add the effect to be obtained from the Keyline dams, the high contour dams and the others dis-cussed in this book.

Such dams, constructed as they are for use whenever required, with their pipe and valve outlets to provide water at the turn of a large tap, will form a tremendous buffer against floods. The conserved water is second only in low cost irrigation to the rain itself. The Australian drought-breaking flooding rains will then find a huge capacity in the farmers' dams ready to offset their intensity and destructive force. The drought will surely have warranted the use of the water of those dams and their capacity will be available for the flood rains.

From geological evidence it is apparent that floods did occur before the farming and grazing practices of our few generations of farmers greatly reduced the capacity of the land to absorb rainfall and retard the sudden flood. It is just as apparent that no rains of recent decades should have caused so much des-truction. In this geological age of lower rainfall and drier conditions, every drop of water, including the rains which now cause our floods, should and could be used in the production of better soil. The soil would probably be better than that which previously existed in the Hunter River Valley.

These remarks are not a suggestion that the Keyline plan will in effect put the clock back 150 years, nor is it suggested that the valleys and streams of that important river watershed will revert to their former state as regards the cleanness of the river flow and the reduction of the quick destroying flood. No! Much more than that is feasible. The whole of the land will rapidly become more fertile and absorbent than it ever was. The heights which the floods reached 150 years ago, which were perhaps much less than those of to-day, would probably not be reached again.

There is no doubt that, at the moment, great flood dangers exist. There is also no doubt that projects of a national character in the construction of many flood control structures would greatly mitigate the danger of the big floods.

Such works cost sums of money that to the ordinary mind are quite fantastic. They require for their finance a toll on the whole of the community. They cover with water large areas of very valuable land.

From a practical business point of view, where is the flood control problem, or any other problem for that matter, if a highly profitable solution is found!

Against the Keyline picture of almost absolute control, we have the ever-present menace of the big flood with something much more than a possibility that a flood larger than the previous worst one could occur at any time with little warning. The only other hope of protection, which lies in the very remote future, is the construction of fabulously costly government-projected flood control dams. If and when sufficient of those are constructed they would not have as great a combined water storage capacity as that which can be had at very little cost in the soil itself by Keyline absorption-fertility.

The reason why soil erosion control or soil conservation has not been accepted by a very large percentage of landowners is simply that such matters are not always good business. Too often it is something to be attempted reluctantly and postponed very easily. The approach is negative, the cost real, and the profit remote.

The phrase "Prevent erosion and save the soil that is left" lacks inspiration. Why not, say the farmer and grazier, forget erosion? Instead, build better soil structure, improve soil fertility, make, manufacture and create deeper, more fertile soil just by providing soil with the capacity to absorb fertility. If a sheet eroded area or an erosion gully is in the path of the better soil drive, convert it; engulf it in the waves of fertility.

If a shire council or the main roads' board is causing large quantities of water to be diverted on to the farmer's land, thereby causing destruction, diffuse it, disperse it, absorb and conserve it in dams. It may be dirty water, but it is water. It is the greatest factor, in the average Australian farmer's mind, in fertile soil development and better yield.

The failure generally to treat agriculture in its entirety by sectionalising and subsectionalising too much with inadequate means of proper co-ordination has led to a completely unnatural and artificial basic approach to land matters. The soil has been lost looking for the crop. The land is being lost while only 80 or 100mm of topsoil is used. Improving and progressively increasing the depth of the soil is the first basis of any permanent yield improvement. Any and all other means of improvement may then logically follow.

Absorption-fertility is real fertility. It is not doctored or drugged soil.

The Keyline plan is complete as a general or basic guide for land development, and has been expanded to meet every arrangement of land and people.

Every method of agriculture which we have used is constantly being critically examined both in Australia and overseas to determine whether it gets its result by extracting fertility or whether it conforms to Keyline's conception of ever-increasing fertility by absorption.

Many new ideas and techniques that were indicated by the general course of the development of Keyline have been fully tested and proved. They include such items as pest control, pasture management, special sowing methods and cheaper and more effective means for soil testing. New methods in the use of fertiliser and trace elements are showing great promise.

Very interesting results of various weed treatments and their effect on soil and pasture have been noted. Some of those weeds are likely to be of great importance and value in rapidly improving very poor soils.

I have no doubt that with the emphasis on absorption-fertility as much as on production, farmers and graziers will find many new and better ways of contributing greater life and value to their Fertile Soil.

Aerial view of Nevallan homestead, with high contour dam in the top left-hand corner of the picture. This dam, which is filled by drains, collects its water from the roads and yards of the homestead area. It has 102 mm outlets through the end walls.

Sons, Ken (left) and Neville, in a newly planted strip of tallow-wood trees. Trees were 152 mm high when planted in the spring and the picture was taken in the following autumn. The strip contains about 1000 trees (vide Chapter 8). Tallow-woods are not indigenous to the district, but are doing well.

PART TWO

CHAPTER ONE

How dry are we?

THERE MUST BE FEW AUSTRALIANS nowadays who do not know that they live in the world's driest continent. Over the last few years that water fact of life has had very wide public mention. Quite a few people, some of them well known, have used the fact as a prelude to some rather startling statements about our "grave shortage of water".

Frequently the rainfull and water runoff records of the United States, which is a country of similar size — eight million square kilometres (three million square miles) — have been quoted against our own lesser figures to drive home the facts. But how many people realise the much more important fact which those oft-quoted figures quite clearly disclose, namely that Australians are much better off for water in Australia than Americans are in the United States?

On the basis of average annual rainfall of 400mm (16ins) in Australia and 740mm (29ins) in America our population of 14 millions would appear to be several times better off than America's 216.5 millions. Against that comparison it could be argued that a considerable part of the rain on this continent falls in dry and very sparsely populated areas and that a great deal of water is lost by evaporation without doing much good or promoting any runoff. Then it becomes very apparent that Australians do not really live in "Dry Australia" but that the vast majority of its people live in the parts of the continent which enjoy the benefit of an average annual rainfall at least equivalent to that of America.

Again, on the basis of runoff water Australia's average is reputed to be 38mm (1½ins) a year against America's 230mm (9ins), so Australians still have a considerable advantage on the same per capita basis. But how much greater is the runoff from the part of Australia where over 95 per cent of our agricultural production is won? It thus appears that Australians are, relatively, very well off for water.

There are, however, two aspects of the discussion on agricultural water where Australia and America do reach a common level: (1) The worshipful respect of both to the Idol of the Big Dam for the supply of vastly larger and more valuable water resources which could be used for the same purposes of irrigation, and which belong on all their farm lands, grazing properties and forested areas. That last great water resource is treated almost as if it didn't exist and certainly as though it were of insignificant value.

Cost of the irrigation hectare

In any field of endeavor where a new asset is to be built up, the various processes for doing it are examined with the utmost care to find the best means of producing it at the lowest cost. But when the new asset to be produced is the "irrigation-hectare" the ordinary business approach is outside the bounds of public thinking, or so it would seem. But the irrigation-hectare, like most other things, has a "value" and since it is a production unit, so the value of it is closely related to what it will produce.

The market price of the irrigation-hectare is determined accordingly. And so its real value is the price which a farmer can afford to pay for it and make a living out of it.

But what is the comparative cost and value position? As far as can be determined the irrigation-hectare produced by large government undertakings costs much more to produce than it is worth when judged on ordinary business standards. The cost appeared to be upwards of $2500 per irrigation-hectare but no big scheme has been built since to give comparative updated costs. (Values throughout this book in Australian currency.)

And what is a hectare of irrigation land worth? Here in Australia its value varies according to where it is and for what it is used. If its use is for the general production of wheat, sheep and wool or for beef breeding then the value would be about $900 a hectare ($160 an acre). However there are some newer intensive grazing systems for which irrigation pasture could be used and where the value of the irrigation hectare could be higher. For the production of milk its value could more than double to $1000 a hectare. Its value could double again were it to be used for horticulture. And in special but very limited circumstances it will even be worth as much as it cost the government to produce.

If those values are at all representative then it follows that land is economical only for the production of crops with a high value a hectare return such as those mentioned last and vegetable, orchard and vine crops. But such crops may quickly reach the stage of overproduction and therefore growing them is often controlled in some measure. Rice growing in the Victoria-New South Wales irrigation areas is a notable example of that type of restriction when it is applied by government authority, while another control is the production of certain vegetable and fruits under contract with food processors. If a farmer produces above or outside a contract he could be left with his crop unsold or may have to sacrifice it for far less than it cost to grow. There is no shortage of those kinds of home-grown agricultural products in this country.

DISECONOMIES OF LARGE SCALE PROJECTS

From those facts it can be seen that any rapid extensions of the present government irrigation schemes would have to be used for those items of production for

HILLSIDE IRRIGATION

PLATE 12

Water flows from the lock-pipe system of a farm dam at the rate of 1400 m^3 (300,000 gallons) an hour into an irrigation channel. This channel is designed to link up with the lock-pipe releases from the second, third and this fourth dam of the creek chain from which it extends another 0.9 km to a point near the fence on the eastern boundary. The irrigation channel has a fall of 1 in 300 and Keyline pattern irrigation is at the rate of 2.4 ha an hour.

PLATE 13

A member of a Keyline school places an irrigation flag in position ahead of the one in use. The fence is positioned a minimum distance of 6.4 m (21 ft) above the irrigation channel. Irrigation rate is 2.4 ha an hour. Water use efficiency is high since, even if the water is misjudged and an irrigation flag is left in position too long, the water in excess of the absorption capacity of the soil is trapped by the next lower channel and transported to a dam. Thus only "height-of-water" is lost.

PLATE 14

Irrigating at the rate of 2.8 ha an hour on some of the steepest land on the property. Rate of flow is 1600 m^3 (350,000 gallons) an hour. Keyline pattern controls the water spreading.

PLATE 15

Looking up towards the channel during irrigation on hillside land with a fall of only 2 in 25. Irrigation rate is 2.8 ha an hour.

PLATE 16

Early scene of the flatter of the undulating land of Yobarnie, showing an irrigation channel in the foreground. It carries 1400 m³ (300,000 gallons) of water an hour on a fall of 1 in 300; irrigation rate is 2.4 ha an hour controlled by irrigation flags. Width at ground level is 7676 mm. This area was later subdivided into small paddocks. A fence now follows the irrigation channel a minimum distance of 6.4 m above it. The up and down fences are on the lines of maximum fall. The lower line is part of our main diversion channel.

RRIGATION AND DIVERSION CHANNELS FOR HILLSIDE IRRIGATION

PLATE 17

The yearly trimming of an irrigation channel with a special delver, including speader bars which moves the earth away from the edge of the channel. A farm tractor equipped with dual wheels has proved to be most suitable for all work on irrigation land.

PLATE 18

A newly constructed diversion channel in medium undulating country. The channel has a fall of 1 in 300 and collects rainfall runoff from the hills above it to fill a farm dam for irrigation.

PLATE 19

A diversion channel of several times greater capacity than that pictured on the previous page. The channel has a similar fall but is much longer. Both diversion channels were constructed by sidecasting with a large bulldozer equipped with a "back-ripper" and an "angle-and-tilt" blade.

PLATE 20

Closeup of an irrigation channel for hillside irrigation. The channel is 2 m wide, has a fall of 1 in 300 and a capacity of 800 m³ (400,000 gallons) an hour. It provides for an irrigation rate of 2 ha an hour which is the maximum for one man control for Keyline pattern irrigation. The marker peg on the left and lower side of the channel indicates the correct position for the construction of the channel in relation to the pegged line.

PLATE 21

The first test flow of water in a flat-land irrigation channel designed to carry up to 9,000 m³ (2,000,000 gallons) an hour; this present flow is 1,800 m³ (400,000 gallons). Flood flow irrigation channels may have a slight fall but generally should be dead flat.

PLATE 22

Keyline cultivation for soil and pasture development. The rough appearance of the third and final cultivation for soil improvement as described in Chapter 6. This rough-finish cultivation provides the drastic soil aeration found to be so advantageous for rapid soil and pasture improvement.

which Australia has increasing overseas demand. And apart from the possible gradual rise in sales of the high value a hectare crops any new large irrigation districts would need to be used for the lower value a hectare returns of general production. So any new irrigation-hectares would still cost very much more than they are worth.

Referring again to the cost of producing irrigation land by government authority, it is very difficult to obtain clear-cut information on such matters, but the figures quoted are on the basis of information on the cost of a storage plus the cost of reticulation structures and divided by the number of hectares which are serviced by the irrigation waters. When it comes to obtaining information on dual or multi-purpose dams where both irrigation and power generation are included, the information is even more confusing. It often seems to the inquirer seeking to know the cost of electric power that it is somewhat higher than it would be from a thermal power station, but "look at the free irrigation water supplied", is the typical reply to such a suggestion. But if the cost of producing irrigation land is sought it is found to be very expensive.

If the irrigation-hectare is not worth what it costs to produce then who pays or who loses the difference between cost and value? A good question. The ordinary taxpayer has the rather naive notion that the farmer benefits so the farmer pays. But the answer is that the ordinary taxpayers are the ones who pay now, and who have the best possible prospect of continuing to pay for the undertakings.

The purpose of these comments on the large "government dam and irrigation district projects" is to show by comparison with them the value of farm waters. If a full scale development of farm waters took place for no other purpose than the production of more irrigation-hectares, then they would have two great advantages over all the large scale projects of government developments.

First, there is vastly more water to deal with and so to produce more irrigation-hectares than from all other sources combined and second, the cost of producing such valuable irrigation land would be very much lower (and costless to the taxpayer by comparison). However the development of farm water resources on the basis discussed throughout this book is on a broader scope than merely the production of irrigation land, as the objective is the planned development of all the resources of agricultural land.

Some illustrations may serve to show the sheer vastness of this great neglected resource. Suppose for instance that Australian soils covering one third of the continent could have their fertility improved and their soil deepened a little, as is described later. The soil would then be able to take in more rainfall and use it effectively. If only an additional 51mm of rain a year were involved, it would be equivalent to over 120,000 million cubic metres (100 million acre feet) of water, and of the cheapest "irrigation" of all. Further, it would constitute an improvement in agriculture never before paralleled.

Does it not appear completely illogical in the first place to do anything about storing water in "big" dams for irrigation at such high cost and at the same time neglect the benefits of soil and landscape improvement and tremendously increased production, which are available so cheaply by simply improving on the use from rainfall where it falls? By comparison now the Snowy Scheme is complete less than 2.4,000 million cubic metres (two million acre feet) a year is added to the waters of the Murray and Murrumbidgee Rivers for irrigation. Australia'a agricultural land of say one third of the country would have an average annual runoff of well over 120,000 million cubic metres (100 million acre feet). So there is more runoff water than in a hundred Snowy schemes. And the spear-grass and brigalow belts of Queensland's vast farm water resources would surely supply enough runoff for many more. Practically all that water falls as rain on farm and grazing lands.

Before delving into the cost position of irrigation hectares from farm waters it would be as well to look for reasons for the over-emphasis of the one and the neglect of the other. Why should government projects completely dominate public policy when the objective is to improve aggregate agricultural production? There are probably many reasons. Among them undoubtedly are the spectacle, the size and the glamor of the undertakings, which make a powerful appeal. At this stage of our national advancement who could argue effectively against such things with the outstanding example of the towns and the landscapes of the older irrigation areas created out of water and near desert? And added to their spectacle are all the conveniences of city life and a bustling industrial complex starting with the factories for processing the product of the irrigation land. For centuries has not a crowning achievement of civilisation been in making the "desert bloom"?

Sectional interests which could greatly benefit from such projects are continually demanding that "something should be done about developing the limited water resources NOW". Since there is no doubt that money follows the water, why not? Of course what is meant by the demand that something be done about the water now is that governments have to be persuaded to do something about it. And how? By arranging that all the ordinary taxpayers of the state or nation be made to pay for something which will directly benefit just a few farmers and the local business communities. Rarely is such an undertaking sound enough from a business point of view for the farmers and business men to get together and finance it themselves or they would do so.

Is the landholder winning or losing?

If the object of an irrigation water policy is to benefit as many farmers as possible or to increase production as much as possible with the money available, then the government dam and irrigation project is most assuredly *not* the way to do it. In fact the rapid extension of such schemes could have the opposite effect of ensuring that the vast majority of farmers cannot get any assistance with their own individual water developments. And if a few farmers get benefits why should not all of them? Despite that it seems almost all farmers tend to favor and support the new schemes — when they should be the chief objectors.

Big dams are big business and big business always has much to say.

On another view there is a special hazard in the concentration of the high-priced agricultural products of irrigation land which is necessarily a feature of government projects. A drop in market prices can be disastrous for the farmers and all local workers and businesses which depend on them.

Not all government irrigation and agricultural water supply schemes are as disproportionately costly. But invariably those which are considered good business propositions are the lesser projects — and those are the ones which the public does not hear about. Often such fine lesser-projects or starting-off schemes are good business to the extent that they would be satisfactory for private capital to support. But all too often the demand for extension or enlargements succeeds, and the reasonable relationship between costs and values disappears. Bigness itself is too often the trap. Often it appears that it matters not how illogical a water scheme may be, so long as it is "big" it will command wide support.

All of those projects which could not be regarded as reasonable business propositions are justified in various arguments, the first being their outstanding importance in national development — and that seems to apply to every single project without exception — and the last one being that in the final analysis the profits of all who benefit are going to be taxed so that ultimately the money spent is returned.

Later reference will be made to the water use efficiency of on-the-farm irrigation developments compared with the large government project. It is said the necessary factor of reliability or complete safety of water supply in the government irrigation schemes entails a tremendous cost in lost water, whereas with farm water development that factor of complete reliability can never be a reasonable basis of planning.

The totally different philosophies of the two developments become much more significant when the runoff figures are considered. The government project can be concerned only with runoff water which has already reached the major streams and, except for the one high mountain development, the Snowy Scheme, the water which is to be stored has already travelled great distances. As with the great artificial supply channels, so with the natural channels of the rivers in earth — the water travel has already cost much in water losses. And as the total runoff is probably somewhat less than 9 percent of total rainfall the loss must be considered a serious one from a national water use efficiency view. Again, it appears that if every drop of runoff were stored in big dams on the rivers, the quantity of water would still not approach that available on the farm and grazing land. It could be also asked; if total runoff is 9 percent what has happened to the remaining 91 percent? Water is lost all the way along the line — everywhere — and the farther it travels the more of it is lost.

All the water which runs off the farms does not reach the river.

CHAPTER TWO

Limitless scope for farm irrigation

OBVIOUSLY THE ONE PLACE where the highest efficiency in water use can be achieved is on the land where it falls as rain. Therefore for agricultural purposes the water should first be used on the land where it falls or as close to it as is possible.

When water use efficiency is discussed it should be clearly realised that the matter of efficiency in water use is purely relative. Likewise so is waste of water. It has nothing in common with the efficiency qualities of industrial processes or of "cutting the coat to suit the cloth". Water is a moving factor, and it continually moves on. That can account for the seeming paradox which emerges on many occasions. It often happens that the more water is used the more water is available to be used, and the sure way to really waste water is not to use it. Country people may sometimes be concerned about the extravagant water habits of city dwellers. But what does it amount to? No matter how frugal was their use of water none of it would be available to the country and for agriculture. And so long as city water does not run short to the stage of water restrictions then it is better that the water be kept moving. It is quite impossible to know even when water is actually being wasted without at the same time knowing when it is next going to rain and how much. There just has to be what is called "waste of water" so why not waste to the ocean the kind that would be the most expensive to hold.

The cost of the farm-developed irrigation-hectare produced as part of the Keyline development of land is met now by the owner of the land and that first fact should insure it is not going to cost more than it is worth. Because of the nature of such water developments and their position as part of a general planned improvement program, the producing of more irrigation land will march step by step with other parts of the improvement program. Moreover, because of the progressive nature of the farm water developments themselves, the farmer is able to continue a program, change it or abandon part or all of it, according to results.

The cost of economically producing the irrigation-hectare from farm water resources which have to be completely developed may range from $250 a hectare upwards.

In many cases farm irrigation from water which was stored for the purpose in a dam, may be considered for its safety or insurance value alone. That type of approach is referred to as "supplemental" irrigation, and it usually constitutes a standby reserve to ensure the production of a special crop, and in case of a dry spell. Generally spray irrigation now serves that lesser purpose, and as the aim is to equip the area as cheaply as possible, irrigation is planned which will water the specific area in the same time as the irrigation cycle.

Such slow rate of irrigating greatly reduces equipment costs in a smaller pump, smaller main lines and spray lines, but irrigation is much more costly in man hours.

That type of irrigation is not a part of the wide and proper development of water resources, as it insures only against losses, but does not assure the production of extra profit. Certainly there is the exception where such a project is for the small area production of high value crops. Schemes of that type may cost from under $700 to over $1400 a hectare. They have a place on some farms, and although the cost an irrigated hectare is low when compared with the hectare of the large public scheme, they are not of the class which should in the future play other than a minor role in any major drive to develop farm water resources for widespread irrigation. Any drive in that direction would need to ensure the larger percentage of the new irrigation land is won at low cost, and that the irrigation procedures for applying the water to the land be as economic as possible in manpower costs so general production can be increased and not rely on a selected variety of high priced crops.

Cost of transport

The cost of water transport in water lost in large government supply channels has been emphasised. There is another matter of transport which accounts for a seeming paradox in cost comparison as between the "big" dam and the farm dam. Generally in industrial processes the greater the production the less the unit cost, but no so with water and irrigation. The "big" dam, which from appearances, should provide storage for water at lesser cost a unit than the farm irrigation dam. But it rarely does so. One reason appears to be the far higher cost a cubic metre of earth for the wall of the "big" dam which is usually over 10 times higher than the cost for the earth placed in the farm dam. And the cost of transporting the material is a large factor in the comparison. Whereas on the farm, a dam site to be used has to have good earth for wall construction at, or usually within, the dam site, a suitable site for a "big" dam has no such favorable feature. The materials have to be much more carefully selected and invariably carried far greater distances. Concrete walls for the structures are much more costly.

There is another significant comparison of the two means of securing irrigation land. In the "big" project the land on which the water is to be used is carefully selected for its suitability for irrigation. Because a free draining soil is a standard first requirement the loss of water in the distributing channels is not lessened. The land to be used for irrigation developed on the farms is also selected, but because of the confined area the selection should not be made on the same basis but rather as to the position of the land in relationship to the water for the cheapest application and with a realisation that if the soil and the landscape can be improved then it can be improved much faster with the aid of irrigation.

The way to improve agricultural land and production most cheaply and rapidly, and very importantly, to improve the income of the farmers and graziers, is by the further economical development of present holdings and not by the too rapid bringing-in of new land. The best land is that which is now producing profitably but which, by and large, is capable of being very significantly improved.

The development of farm water resources would create a collectively vast area of irrigation land spread widely throughout all the farming and grazing districts. It would not be dangerously concentrated in the one place or concentrated on the one class of production. It would thus create the balanced type of increase in all fields of agricultural production which would best serve the nation's progress.

The "big" dam will invariably be the best means for providing the water for the large centres of populations where the value of the water is high. But irrigation water from government sources costs the user from 0.12 to 0.24 cents a cubic metre ($1.50-$3.00 per acre foot) while city water costs 20-40 times as much. Therefore if the expanding population of a big industrial city wants the water of an irrigation dam, who is most likely to get the water? The irrigating farmer? Not likely!

A comparison of procedures

Irrigation procedures for both government irrigation district and farm developed water can now be considered and related. It will be shown that the relative size of some aspects of the two ways of handling agricultural water is not always in favor of the government scheme.

One factor to be borne in mind is that once the water is on the farm and the actual irrigation is under way, the significance of the "size" of the government water scheme disappears. The water may have been stored in one of the largest structures man can make; it may have travelled many hundreds of kilometres in huge supply channels; been diverted by ingenious and costly control gates to several smaller and still smaller channels. But only when it arrives on the farm does the real irrigation project start. The size of the irrigation project is governed by the volume of the flow of that final stream of water which the farmer then diverts onto his land for irrigation.

How big is the government irrigation project now? The general answer is that as the water is sold it is measured. The measuring device is the dethridge wheel which measures flowing water up to 0.15 metres a second (five cubic feet a second). Therefore the maximum size of the irrigation stream is now $540m^3$ an hour (112,500 gallons an hour), or usually considerably less. An average flow of water of about $370m^3$ an hour (80,000 gallons an hour) is then controlled by one man using the special features for the purpose which were provided by the various land preparation methods.

How large a flow is $370m^3$ an hour (80,000 gallons) of water an hour? In an excavated irrigation channel as used for hillside irrigation by the Keyline Pattern method, a $370m^3$ flow appears and acts as a stream too small for working with, as the method and the farmer could handle four times as much water and thus irrigate faster and at lower cost. As a stream flowing in the type of channel used in flatter government irrigation areas and with a very flat fall of, say, 1 in 2000, it appears to be a large and impressive volume of water until it is diverted to irrigate the land, when it again becomes unimpressive.

As an irrigation stream it will water 0.56ha (1.4 acres) in one hour so long as it is so controlled that only 51mm of water soak into the soil.

I believe those irrigation rates are far too slow and too costly in manpower. However, the maximum size of the irrigation area which should be serviced by that stream is 22ha (though that opinion may not be accepted by irrigators without Keyline irrigation experience).

Border check irrigation

Border check irrigation, also referred to as "border bay" and "border strip" is used widely in government irrigation districts. Water is released from an irrigation channel constructed generally across the fall of the land, into bays formed by small earth banks, or borders, which extend from the channel at right angles and down the fall of the land. The object is to advance a sheet-like stream of water down the strip of land between the low banks and cause the required amount of water to soak into the soil.

The borders are generally about 10m apart and the bays formed by them may be up to about 140m long. Border check irrigation is used for gently sloping land and up to 1 in 30 or 3 percent, but it could be used for pasture on slopes somewhat steeper on stable soils and on up to five or six percent slopes by reducing the bay width to 4.8m.

Considerable land preparation is necessary involving instrument levelling and pegging-out to mark the area for cuts and fills, several cultivations to loosen up the soil and supervision of the earthmoving to smooth and level off the whole area. The small banks are formed in two moves, first a small grader with straight blade traverses the whole field working parallel to the irrigation channel. At the crossing point in turn of each bank line, which has been clearly pegged, the blade is raised to release its load of earth. At the completion of that task the earth for all the banks is in place, forming uneven lines of banks. Second, a "crowder" traverses each of the banks lines, crowding them into neat even banks about 254mm high.

The irrigation channel, usually called a ditch, may be equipped with small permanent watergates through which the water is released into the bays, or the bank of the ditch may be opened with a shovel and closed again with earth when irrigation of the bay is completed. Several bays may be receiving irrigation water at the one time. The irrigation ditch receives its water from the "head ditch" which is the main channel on the farm and where the water is measured. Land preparation costs may average $200 a hectare, but on occasions be either considerably higher or lower. Border check irrigation is used for the growing of a wide variety of crops, including orchards, but it is especially suited for small grain crops and pasture production.

But if a much greater flow of water were available as from a farm dam, for bigger areas, then according to the land slope and the water resources available the method of border checks could be abandoned in favor of one of two much faster and cheaper methods.

First, if the steepest sloping land to be irrigated is 3 percent, or 1 in 33, and water is available for upwards of 32ha of irrigation then Keyline flood-flow (about which much will be said later) could replace the border check method and would be suitable for growing the same range of crops. The design can provide for a water flow rate of $2250m^3$ (500,000 gallons) an hour to complete 32ha of irrigation in an eight hour day.

Second, if the flattest slope is 3 percent and the range of slope is from 3 percent to 8 percent or steeper, the border check method can be dropped in favor of Keyline pattern irrigation. For a similar area of 32ha the designed rate of flow is from 1100 to $1400m^3$ (250,000 to 320,000 gallons) an hour and the irrigation of the area can be completed in two days or less. Unlike border check irrigation on that range of slope, which is for pasture irrigation only, Keyline hillside irrigation is also suitable for a wide range of crops.

For both those replacement methods the water is supplied from new farm dams by means either of a lockpipe beneath the wall of the dam or by large-volume low-head pumping, according to the circumstances.

Contour bay irrigation

Contour bay irrigation is also widely used in government irrigation projects. It is most satisfactorily employed where slopes are somewhat flatter with, say, a fall of 1 in 1000.

Unlike border check the irrigation water supply channel preferably falls directly down the gentle fall of the land. The irrigation bays are again formed by earth banks and they extend outward and at roughly right angles from the irrigation channels and are constructed on a true contour. The bays are closed in at the ends lying away from the irrigation channel with a similar sized earth bank.

For irrigating, water is diverted through water gates into the first or highest bay until all the land of the bay is covered or flooded with standing water. Then the gates from the irrigation channel are opened into the second irrigation bay, the gates to the first bays are closed and gates in the lower bank of the first bay are opened. Thus the full stream from the irrigation channel and the water remaining in the first bay flows into the second bay. The irrigation channel itself is banked across at intervals to hold the water and prevent it from bypassing the open bay gates. The banks across the irrigation channel are also equipped with watergates.

Land preparation costs are usually considerably lower than for the border check method. The irrigation channel may be wider and shallower. The banks for the bays are much larger. The distance apart of the contour banks is considered satisfactory on a cross fall of 76mm. On a 1 in 1000 slope and with a cross fall of 76mm the banks would be 76m apart. The banks themselves, although larger, are much more cheaply constructed by the sidecasting method. The shallow ditch left is usually untouched and not "stopped".

Contour bay irrigation is suitable for growing a very wide variety of crops, including orchard and pasture. On suitable soils rice is grown by this method with only slight variations.

Contour bay irrigation from farm-developed water resources

Contour bay irrigation from farm developed water resources follows in detail the above layout and procedures. Improvements come mainly from increasing the rate of flow of the water supply, as the contour bay system, with only the necessary alteration of enlarging the irrigation supply channel, is capable of using large flows of water quite effectively. An alternative procedure is Keyline flood-flow irrigation, where even larger flows can be controlled by the one man, where land preparation costs are still lower and, most importantly, where drainage, often a problem of contour bay irrigation, is not a problem.

Furrow irrigation

Furrow irrigation is the most widely used system for the production of row crops and occasionally for orchard irrigation. On uniform and slightly sloping land the irrigation channel lies across the slope of the land and the furrows fall directly down the slope and at right angles to the irrigation ditch. The crops are grown on the small banks of soil between the furrows. Water is delivered via small pipes through the ditch bank or more conveniently by specially formed syphon tubes over the bank, to many furrows at a time. The small individual streams follow the furrows downward and thus water the crop growing between them. In circumstances where slopes are too steep for that layout the irrigation ditch may be on or very near a true contour and the furrow positions will be pegged in with the aid of a level instrument to fall at a selected and slight grade. One furrow of each five or six may be thus placed, and those intermediately spaced will be parallel to the adjacently marked furrows.

Spray irrigation

Spray irrigation had been little used in government irrigation districts until the last 20-odd years. While there has been a notable increase, the land watered by spray irrigation is still only a very small percentage of the total. However in the more humid areas and for supplemental irrigation its use has become more widespread. Certain advantages are claimed for spray irrigation, namely:

1. Uniform and closely regulated quantities of water can be applied on any type of soil, particularly porous and sandy soils which do not hold water well.
2. There is no land preparation cost involved, and with an absence of channels there is no "wasted" land.
3. It lessens the need for cultivation to control weed growth, as against furrow irrigation.
4. Small area irrigation from small flows in streams or from little ponds is possible for certain high value crops.
5. Soil erosion can be controlled on steeper hillsides.

The principal limitations to spray irrigation are:

1. Initial cost is extremely high.
2. It is unsuitable in windy conditions since the pattern of the spray is disturbed and watering is uneven.
3. Water must be free of rubbish and sand which choke up the spray nozzles.
4. Power costs are extremely high because of the need to provide continuous pressure.

There have been continuing big improvements in spray irrigation equipment over recent years, and as there will always be circumstances where water is available and other less costly methods will be impracticable, srpay irrigation will continue to play its role both in government irrigation and from farm water resources. However, since there is a plentiful supply of literature dealing with the

various types of equipment and the differing procedures for spray irrigation
they will not be dealt with here.

Throughout these discussions a comparative analysis emerges on the efficacy
of the two ways of developing irrigation land: (1) By government "big" dam
and irrigation area project and (2) by the logical development of farm water
resources.

The "big" dam and the great supply channels and the tight concentrations
of irrigation farms in the irrigation districts are the impressive, even overawing,
features of the former. But from every possible aspect, the extension of irrigated
land from farm and grazing land water resources is now the best way for maximum national benefit, and the way to improve incomes for the greatest numbers
of farmers and graziers.

The related size of the two when the water actually comes into use on the
farms is very much in favor of the latter, and after all placing water into the
control of the individual farmer is the object of both.

In the government scheme large flows cannot be kept available for the irrigator for just when he happens to need or wants to use the water. In fact the opposite is the case; the whole working and management of the farm is usually
ordered by the times when the farmer is allowed to take water. The rate of
supply has to be limited, not by what the irrigator may be able to use, but by
the necessity for the water authority to keep the supply channels down to a reasonable size, and to be able to service as many irrigators as possible from channels
of limited capacity. Therefore the rate at which the irrigator can water his land
is not governed by his capacity to do so, but by rate of the available supply and
that also governs his manpower costs in irrigating.

Limitless possibilities of farm-water irrigation

Water from a farm irrigation dam, on the other hand, is not affected by such
limitations nor need it be restricted in the rate of flow of the water by other
than the practical considerations of the capacity to use the water and by the size
of the dam and the related area of the irrigation paddock. If the land can be
watered at a rate of flow only similar to that of the government scheme then
that rate is used, but if two, four and 10 times that flow rate is practical, then
the greater flow can be planned for and used.

A government scheme, no matter how large could not supply each of its
farmer irrigators with a flow of $5500m^3$ (one million gallons) an hour. Such
an idea would be considered quite ridiculous, and such a flow completely uncontrollable by any irrigation procedure. But that need not be so on a farm.
That large flow for some farm developments is quite practicable, being simple
to design, economical of construction and fully controllable by one man and as
a normal irrigation stream. It all depends on the particular circumstances, and
when these suit fast and extremely economical irrigation why not design the
work according to those favorable circumstances?

For hillside irrigation, which is not a feature of government irrigation, does
not the land possess especially attractive irrigation features? It is not likely to be
troubled by those serious problems of irrigation districts, waterlogging and poor
drainage. Hillside land, in the very limited field of its use, is spray irrigated at the
very slow rates of from 0.8 to 1.6ha (two up to four acres) a day. But "hillside
irrigation" by the Keyline pattern irrigation system is eminently practical in a
very wide set of conditions and at watering rates of 20ha an eight hour day and
with only one-man control. Also one former bugbear of hillside irrigation, the

danger of soil erosion, is not, with this system, a factor for more than passing consideration.

The water available for irrigation developments, which is associated with hillside land, constitutes an enormous untapped resource spread widely throughout the farm and grazing lands.

It has been seen that every method of applying water to land which is in wide use in irrigation districts can be used with at least equal advantage on farms and from their own water resources. But further than that, because the flow rates of water on those ordinary farms are not so restricted, they allow every one of these methods to be greatly improved upon. This is particularly the case with lower costs of labor when more hectares can be covered each day by one man.

Not all farms have water resources which can be developed for irrigation, and some with sufficient water may not be able to use it because the topography of the land precludes it, or because the available soils are not suitable for dam building. Problems of materials will gradually be resolved, and particularly so when it becomes the serious business of some government departments to conduct tests and to make wide experiments conducted on average farms, and with sufficient money to support the work and solve the problems.

There appears to be no basis to the often repeated statement that Australia is seriously backward in the development of her water resources, meaning of course the "big" dam type. In fact we seem to be as well advanced in water development of that kind as any other progressive country, and particularly is it true on a population basis. So much so that in future that type of project should be examined much more critically than in the past. It should be acknowledged that in many cases with the "big" dam project (and even with on-the-farm waters as well for that matter), it can be good business to do nothing about the water but just to let it go. There just has to be what we call "waste of water" so why not "waste" the kind which would be the most expensive to hold?

It is now time for governments to encourage by some special measures the development of the on-the-farm water resources. It would be good business for governments if they could promote the development of a very large total area of irrigation land at say one tenth the price a hectare it had been previously paying for it. But even that low cost would be far beyond what any reasonable assistance scheme could involve financially. In the greater proportion of cases a practical lead and some loan finance would be sufficient inducement to persuade farmers to think again about these matters.

But if it is necessary or advisable to offer bounties of various kinds, or tax concessions on the production increase which result from new irrigation lands, as is done to the export of the products of secondary industry, then that would still be a means of creating new irrigation land at a cost much lower than that by any other means possible.

CHAPTER THREE

Droughts, floods and conservation

WHILE IT HAS BEEN STATED that Australia is not short of water on any comparable per capita basis, the incidence of rainfall is so unreliable because of our geographical position on the globe and the attendant prevailing winds and the absence of well-spread high mountain ranges, that a succession of droughts and floods characterises our climate. Whether it be good or bad, this particular climatic feature is the background to all our agricultural pursuits, and it is therefore a principal factor in any planned development of land. Also the now ingrained "conservation complex" remains as a disturbing problem to overcome before all our land can reach its peak of development.

Droughts and floods are invariably coupled together and they are the major parts of the one climate. In the winter of 1965, with no good general rains since the heavy rains and flooding of the previous winter, the greater area of the three eastern States experienced the most severe drought ever recorded. The city of Sydney, then with a population of over two million, recorded 787mm less rainfall for the year than its yearly average of 1912mm. And there are countless farmers and graziers who would welcome any type of rains, including those which cause our major floods. That illustrates one of the great differences between droughts and floods. Whereas droughts are uncontrollable and, I believe, always will be and will always cost a lot of money, floods represent neglected opportunities.

The flood rains of Victoria and New South Wales

The rain which causes the floods in New South Wales and Victoria are of the order of a total of 254mm which usually occur in two series of falls, one following rapidly upon the other. That 254mm of rain represents, at least to the extent of 70 percent of it, "opportunity" rain which could be made good use of if the control structures for water resources existed on most farms where such things would be appropriate. Moreover severe droughts tend to be broken by flooding rains, and that is all to the good as the end of the drought finds on-the-farm storages carrying their lowest reserves of water.

For instance, if I could, at the height of the 1965 drought, have had the option of either 102mm or 254mm of rain, I would have chosen the latter as the irrigation water of our dams was, at that time, largely used up and our capacity to take in water was close to its maximum.

Continuing the supposition of a fall of 254mm of rain; 76mm and most probably nearer 102mm of it would have been absorbed rapidly by the good soil of the 243ha of very dry rain pasture land and a little less by the 202ha of land which we irrigated. By then the two small creeks flowing to the area, which were quite dry, would have started flowing and filling the dams located on them.

After the first 76mm of rain, runoff from the land of the farm itself would start and be trapped by the diversion channels and flow into the two series of interconnected dams which the channels service. The lower of the two diversion channels on Yobarnie, which also diverts the flood flows of the small creek, would start to receive water from the overflow of the lesser creek storage and start the more rapid filling of the first of the next four dams on that channel.

By the time 178-190mm of rain had fallen the lower series with its larger dams would be filled to the last dam of the chain, which would then be close to the point of overflow. In the meantime, of the higher series of dams, four of the six would have filled with their overflow filling the two remaining dams. The continuing flow from the rain so far received would fill all but two high dams, which are above the chain and which are finally filled and later on replenished from a permanent pump and pipeline from the larger of the creek storages. One creek dam would have been long since filled and the full flow of the water of the creek would be returning back to the creek below the dam. There would also be a considerable flow of water leaving the property in one creek which was uneconomical for us to control.

The channels and dams are patrolled on such occasions, and with several dams filled, the overflow points and the lockpipe valves would be opened to reduce the flows in the diversion channel. Creek dam lockpipes may remain open for two days or more to reduce and then to stop the overflow in the spillways. Of the 254mm of rain the runoff from the land of the farm itself would have contributed little to creek flows lower down, but a considerable volume of water from outside the farm would have flowed on through the property.

A day of sunshine following the 254mm of rain would give an appearance to the farm as if it had had a nice shower of rain the day before, except for the fact of all the filled dams and the water still flowing gently in some of the control channels. Those are what have been called the "damaging-flood-rains" of Victoria and New South Wales.

The summer flood rains of Queensland

The summer flood rains of Queensland are an entirely different matter. Here the flood-causing rains are likely to be nearer 508mm than 254mm, and so the control and storage of the major part of those rains on farm and grazing properties is somewhat more than an incidental in farm water resources development.

Flood rains of such great proportions can not be all brought under control at a profit, either privately from the landowner's point of view or nationally from the viewpoint of the taxpayer. The only logical way to handle the situation is that as much of the rain be controlled and stored on the farm and grazing land as is both practical and economical, and let the rest of the water flow to the sea. But such is the appeal to the compassion of our people when the disaster of flood causes houses to be destroyed and personal belongings to be washed away and lives to be lost, that funds to assist the victims of the flood are invariably forthcoming. Calls for massive "flood mitigation" then break out as if by infection for a few weeks.

But if public works for flood mitigation are to be undertaken it would be better that the money be spent in assisting in the farm control of those waters instead of trying to control the water after it has reached the river; as then whatever is done must still be too little, as well as far too late.

Losses from floods and droughts

Floods cause heavy financial losses which are born by an unfortunate few. But the losses are quite minor from the national aspect and particularly so when they are considered against losses caused by drought. While comparatively few farms suffer losses by floods a very great number are greatly benefitted by the same rains. But during drought the position is reversed. Most farms and grazing properties suffer severe losses, some owners even to the point of the loss of their property, while only a few are fortunately placed to benefit from the disaster.

If it were deemed necessary to make up the losses caused by floods to the owners of flooded land, it could be done with little effect on the public fund. But losses from drought are of such a high order as to be quite outside the scope of such assistance. Droughts and flood cost money, but what a difference in price!

Drought cannot be controlled. But for the individual owner of land, by the full development of the water resources of the farm, he may be able to change the order from six fair years and the one of near disaster, to six very good years and then one lean year.

Artificial rainmaking does not appear to be a possible answer to droughts as there have to be suitable clouds on which to work and they are rare in drought times. However if the experiments were directed and showed promise, at creating the cloud when there was suitable moisture in the atmosphere above the dried-out land and the moisture is of much more frequent occurrence, then there may be some prospects of success in this "last hope" direction; or is it also a lost hope?

Soil conservation versus Keyline

The final aspect of the problems has so far not been discussed. Elsewhere I have expressed my views on the futility of the philosophy of soil conservation, its techniques and the manner of its segregation from being a branch of agriculture. But in 1943, I was an enthusiastic soil conservationist, and on taking over the property at Richmond in New South Wales, immediately began to fight the soil erosion which was evident everywhere on this land.

I had built my first earth dam in 1939, studied soil conservation for some years and had construction equipment including bulldozers and gaint scoops on nearby mining and construction jobs. Therefore I considered myself well-equipped for the work of soil-conserving the property, although I had had no direct experience of farming and grazing.

However my brother-in-law James Barnes was well experienced on that side and was going to live on the property and manage the project.

With the start of work it was found that each technique of soil erosion control I knew so well was not effective in improving the fertility of the soil. There seemed also to be no logical starting point anywhere or basis of planning in soil conservation. But as we needed to store water we would have to build dams as a first necessity. As my earlier dam-building experience had been for mining, and with dams larger than ordinary farm dams, it was decided to build our farm dams much bigger than usual and irrigate from them. That gave us a logical new use for the "contour drains" of soil conservation techniques, which was to collect water to fill the dams. The weather at the time was dry and hot and in December 1944 high winds drove a distant small fire into a general bush-fire. We were burned out and my brother-in-law lost his life in the fire.

The project was continued on the basis that we would try to make all the poor soil into good soil, and as we needed all the water available for the purpose we would catch extra rainfall in the soil and store all the runoff for irrigation.

The whole point of this brief recital is that I had changed from a "conservationist" into a "developer".

The problem of changing poor soil into good fertile soil was solved, and the continued irrigation experiments led to new methods for hillside land that were very much faster than our spray systems, and less costly in capital outlay and in manpower.

Gradually the futility of the conservation approach was realised and later on stated. But such is the power of continual publicity and suggestion that we at first considered we were originating new methods of soil and water "conservation". We were doing what all soil conservationists do to this day. We were making the word conservation mean something totally different.

Very early I realised that mistake, and have opposed the whole idea of soil conservation ever since.

In Australia, with its generally poor soil and unpredictable rainfall, the word conservation is an illogical reference which should now give way to the more practical approach to soil and land "development", including as its main planning basis — farm water control.

If the Department of Mines was named the Department of Conservation that would be a practical use of the word, as every land should "conserve" those things which, when used, wasted or lost are gone forever. And doesn't that apply to such things as the base metal, the minerals of every kind and to our oil and natural gas? But to apply the word and the philosophy associated with the word, to soil and to water — the alive and the self-renewing, — if we permit them to be, — seems now as out of place as it would be to describe the world famous Haddon Rig sheep station in northwestern NSW as an outstanding example of sheep conservation.

Modern soil conservation was born of panic and pessimism in the depression days around 1930 in America and had two basic foundations: (1) the most rapid despoiling of soil in the history of civilisation, caused by the failure to adapt the farming traditions of the older lands to the very different conditions of soils and climate of the New World, and (2) the worst and most strange financial depression in all history; a depression in which, in the midst of an overabundance of foods of all kind, organized society was unable to feed and clothe its people.

And so they starved in the richest country on earth.

And the second foundation of the soil conservation drive was the most activating one, as it was used as the means of putting government money into the hands of a great army of unemployed and so to get the economy of the country going again. For instance one United States Government agency was formed to manage and supervise 32 million ha (million acres) of public grazing land. Another was set up to buy land too poor to afford a living for the farmers and the lands turned into wildlife sanctuaries, forest and recreation areas while the dispossessed populations were transferred to better land. The Civilian Conservation Corps at that time employed half a million men. Such were the huge money resources involved.

But soil, water and forests can not merely be conserved and at the same time be made to fulfill the needs of Australia. Soil must be improved, made more fertile and more productive, and housed in an improving landscape which is necessary for its lasting fertility.

All our water resources should be examined for development and use, and strictly on the basis of economics which should determine the course of the work and the priorities for the different field of use. We just cannot afford the waste of our manpower and other resources on uneconomic development. There is far too much to be done! Water must be wasted, so it is best to waste the expensive kind!

Forestry is not conservation, and particularly not in this country, where it is a function of forestry departments to destroy the trees on tens of thousands of hectares of naturally wooded lands, which are covered with worthless timber, to plant new forests of trees for better timbers. That is surely forestry development as it should be.

The earlier soil conservation ideas of land usages according to certain carefully tabulated land classes have presumed that the best which could be done about any soil was to maintain it in its present condition and not to lose it by soil erosion. There is even now no general realisation that all soils can be tremendously improved, or that merely saving poor soil is scarcely worthwhile.

The only sure and economic way to prevent soil erosion is by forgetting it and concentrating on soil and landscape development.

Of the Australian States, New South Wales has had a soil conservation service for a quarter of a century. That service has cost tens of millions of dollars. Yet even now none of its farm projects or any of its service-owned and costly demonstration farms exhibits a proper basis of land planning or design. And why is it so? In my opinion the philosophy of conservation precludes the possibility of any method based on this misconception from being the best way to improve the land. And a great deal of money has been spent less effectively than it could have been.

Planning based on defence always fails. But when it comes to defence against soil erosion it is foredoomed as the objective is not basically worthwhile. Moreover the redundant earthworks left when soil erosion is stopped – and it can be stopped if the money is spent – remains as a positive handicap to the full improvement of the whole landscape of the property.

It is past time now for governments in this country to reconsider the function of the departments of agriculture and the departments within the conservation department – water conservation, forestry and soil conservation – and initiate completely new and more logical policies of planning and development.

CHAPTER FOUR

Keyline flat land irrigation

MY FORMER PROPERTIES, Nevallan and Yobarnie at North Richmond, New South Wales are of a medium undulating character, where the slopes range from 1 in 3 to 1 in 26. It was from that topographical background, plus that of the low fertility clay soils and the completely unreliable incidence of the 660mm average annual rainfall, that the principles of Keyline were developed from work begun in 1943.

While there was no flat land on Nevallan and Yobarnie, we were aware, and it was soon proved, that the general techniques of Keyline would be applicable to the much flatter areas. However where, on these flatter lands, Keyline planning involves water control for low cost irrigation, then considerable alteration in the planning detail and construction procedure is necessary.

The new methods of water control were worked out and first tried at Yobarnie, and it was soon discovered that low cost flood irrigation was practicable on our most gentle slopes.

The principal purpose of this book is to explain this new aspect of the Keyline Plan — the irrigation of flat or nearly flat country. It has been named Keyline flood-flow irrigation.

It embraces first, the accumulation and mobilisation of relatively large quantities of farm water from rainfall runoff, stream flow or from underground, and second, the design and control of those waters in large flow volumes for irrigating the flatter lands.

The Keyline planning on properties where suitable water exists or where a water resource can be developed provides the design for the mobilisation or accumulation of large quantities of water and the design for the control and release onto the land of the water in a volume in "flood" proportions. The water may be from stored rainfall run-off, or from stream flow, or it may be accumulated in a farm reservoir from underground.

The result of planned water control is that one man can irrigate at a rate upwards of 8ha an hour for the production of crops, pasture for sheep and cattle, or for dairy production. That claim has been amply demonstrated in practice.

In the irrigation of undulating land by the system of irrigation which I named Keyline pattern irrigation, an irrigation rate of up to 3.2ha an hour a man is a simple matter, and it has been for many years. But there are certain factors of farm channel construction and water control which impose limits. Faster rates than that with Keyline pattern irrigation on undulating country are not as yet considered widely practical. However, in the flatter land flood-flow irrigation, those same limiting factors do not apply, consequently the rate of irrigation is limited only by the available water and the shape and slopes of the land.

In flood-flow irrigation, and apart from the matters of the water supply, the land preparation construction costs an irrigated hectare do not increase with the greater rate of irrigation, and indeed may be reduced. For instance, if a projected irrigation stream of $2250m^3$ (half a million gallons) an hour could be increased two, three, or four fold, land preparation costs a hectare are proportionately reduced. In the first place one man would be irrigating at a rate of about 4ha an hour. With the increased flow he could irrigate two, three, or four times, as much land an hour and do it almost as easily. In addition, while that is being done, the methods of water control and soil management within the Keyline plan enable the irrigator to also exercise control of the soil's water intake, which is at least equal to and generally better than that obtainable in a more orthodox system irrigating only 0.8 to 1.6ha a day.

The flood-flow system is also the logical application to the flatter lands of the principles and some of the techniques of the Keyline pattern irrigation system which is now in wide use for irrigating undulating lands. And like pattern irrigation it is a fully controlled system of irrigation from water resources available or developed on the farm and the grazing lands or forest areas.

In Keyline pattern irrigation for undulating lands with slopes of 1 in 4 to 1 in 30 or 40 the Keyline pattern of cultivation is designed to spread the irrigation water and so overcome its natural gravitational direction of flow, which would produce a more or less concentrated stream, down the steepest path of the land towards the valley below. But with Keyline flood-flow for the flatter lands, the pattern of cultivating the land is employed for the more intimate spreading of the irrigation water within irrigation bays which are formed by water steering banks.

Now on land of gentle slope even a very large stream of water will spread widely and so dissipate its power of movement which it derives from the large volume of flow. But with steering banks the stream is held within a confined course, thus preserving its positive forward movement. That has to be understood — there is a great work force residing in or resulting from the sheer volume of a large flow of water. The Keyline flood-flow irrigation system is designed to use that work force with the maximum efficiency, both in its effectively spreading the water over the land and in the crucial intake of that water into the soil itself.

The new development, like the earlier Keyline pattern irrigation and every other Keyline technique, is designed primarily to be a part of a complete Keyline plan for the property on which it is to be applied. A Keyline plan is always individual to each particular farm or grazing property; the shape of the land in relationship to the water resources available for development is the deciding factor in the plan layout and in the methods of applying irrigation water to the land.

These privately owned on-the-farm water resources may be developed to provide the water for various systems of irrigation; thus any of the several methods of spray, flow or flood irrigation may be employed according to the particular circumstances existing on the different properties.

To make this descriptive exposition of Keyline flood-flow irrigation fully intelligible, an explanation of the principles, approach and techniques of the basic Keyline plan is provided in the later chapters of this book.

While farm waters have been neglected as we have said, there has been, on the other hand, a notable but relatively small part of this great potential only recently used in the construction of farm dams for supplying water for irrigation. But spray type irrigation has so dominated the thinking in that development that any mention now of "irrigation" from on-the-farm waters is often taken to mean spray irrigation only. There has thus resulted a complete neglect, even an unawareness, of every other method of applying water to the land. The illogical and very unfortunate nature of this emphasis, and the reasons for it, are discussed later.

In considering the various methods of irrigation it should be kept in mind there is not always a clearly discernible area of distinction where the one system ceases to be applicable and another system replaces it. On some occasions, the unique techniques of the one may overlap those of the other. And that could be so where the choice may happen to be between Keyline pattern irrigation and the newer flood-flow system.

But as a principal aim in irrigation should always be the application of the water to the land at the lowest possible cost, and as low manpower cost can be best achieved by increasing the area of land one man is able to control and effectively irrigate in a day, the choice between Keyline pattern and flood-flow irrigation systems will generally be in favor of flood-flow wherever it can be used, because it permits the control of a greater flow of water.

However, the farmer or grazier should make it his business to know something of the intimate techniques of all systems of irrigation so that he may select the one which he will use in the development of the water resources of his own land. Whether irrigation will produce much profit or result in substantial losses will often depend, in the first instance, on just that choice of the one or other of the irrigation procedures.

Although the emphasis in this book is on farm irrigation, I am totally opposed to the general view of farm irrigation as a thing apart, having little to do with the general development of farm and grazing land. It will become abundantly clear before the conclusion of this book, I hope, that the planning for the development of all land must be based on the various "waterlines" of the land itself, and that is so even when the farm water resources are unsuitable for any irrigation developments.

The planning for the development of all land must have as its main object the maximum profitable use of every natural and renewing resource of the land itself, so that where water resources exist their possible use in irrigation is automatically a part of whole-farm planning.

The principal plea of this book is not that farmers and graziers should use the Keyline flood-flow irrigation system for applying their water resources to their land, but that they should study and fully understand their land and its water resources, and develop them where it is appropriate and promises good profits — and then that they should apply those waters to their land by the one of the several methods of irrigation which is the most suitable for their particular circumstances. Keyline flood-flow irrigation should then be the system chosen only where it can satisfy the requirements of being the best, the lowest cost and the highest profit producing system of irrigating specific areas of land.

But because of the particular properties of water and the precisely ordered lines which are necessary for its effective control and use, the various water lines are the basic framework for the whole edifice of land planning.

CHAPTER FIVE

Basis of Keyline land planning

TO PLAN THE DEVELOPMENT and the improvement of agricultural land, or of land for an agricultural purpose, it is necessary to know just what is wanted. Is it that which (1) eventually shows the most profit and (2) will be the best possible in the circumstances? Or should there be some other important objective? Then (1) and (2) must come into the picture somewhere.

The planner must know what the particular area is capable of becoming: In other words there must be a capacity for assessing the development potential of the particular area. What constitutes potentials of land? How can the potentials be discovered and assessed? Not until satisfactory answers to those two questions are known is the planner in a position to plan.

One class of the potentials of land are those things which largely determine the value of land, namely the price it will fetch on the open market at any particular time. Some of those potentials do not depend directly on the specified area itself but relate to outside influences, such as the market prices for the products which will be grown on the land. Other values of a similar nature relate to the locality of the farm; to the distance from a town and whether or not the town is progressive and expanding; to the conditions of public roads, and to the community facilities available to the farm or grazing property.

The other class of potential, and the one which concerns assessments for improvements, is influenced by such factors as the type of land, whether flat, undulating or hilly, the soil and the rainfall, and other aspects of climate.

The maximum of the potentials of any piece of land depends to a great degree on the capacity of its soil for improvement, and ultimately the principal means of improving the soil lies in man's own capacity to control the water and air within the soil.

Much new land in Australia is brought into agricultural production by first clearing it of scrub and timber. If it is to be used for pasture production, grasses and clovers will then be grown by any of several methods which, with the kinds of seed sown, will usually follow the general pattern of the district. But those improvements, plus the usual subdivision fences, are not ones which use the full resources of the land. Indeed in the ordinary means of "bringing in land" these potentials may be little used, since they seldom, even to moderate advantage, use the water available.

The geography of Keyline

Before discussing the assessment of the water resources of land, as a principal means of soil and land development, the Keyline classification of the shapes of land will be presented, since those features of the land are somewhat dominating in their influence on water movement and on the means for its control and use. Even so, it will soon be seen that land cannot be classified with continuous reference to water, the subjects of land shape and water being nearly inseparable.

To begin with, the settled shape of agricultural land, with its rounded and smoothed out ridges and valleys, is a water shape. No matter what its basic geology, the finishing touches to the land shape have been imparted by the climatic elements, especially rainfall and flowing water.

The two principal "lines" of any agricultural land are water lines. They are, first, the stream courses of every size and, second, the ridges which divide the rainfall water which falls and flows to the valleys and streams on either side of them.

Primary land forms

The stream courses naturally follow the bottoms of valleys which flowing water have formed and, just as stream courses have other streams flowing into them, so also do valleys have other valleys flowing to and joining them: all except one type of valley. That is the "primary" valley of the farm and grazing lands, and the first land shape of Keyline. The valley starts or heads in a ridge and sheds its runoff rainfall water to the stream course below. Water is thus divided by ridges to concentrate as streams in valleys.

The two "shapes" or "forms" of land then are ridge shapes and valley shapes. Practically all agricultural land holdings are either ridge or valley shaped, part of one or both of those shapes or contains one or more of either or both of them.

A good starting point from which to consider the primary land forms which are significant to agriculture, is the juncture of two watercourses. The watercourses may be permanent streams, perennial streams or intermittent watercourses. From such a junction and looking upstream, the land between them will contain the line of a water-divide or a water parting. Runoff water provided by rainfall from the land between the two watercourses will be divided by that line and flow to its respective stream. Again, along that waterline divide, water falling as rain only millimetres away on either side will flow in opposite directions. This water-divide line lies along a ridge shape which continues until it joins another similar ridge, which also is a water-divide line or watershed line separating the waters of the two streams from yet others. (See figure 1.) That is the main ridge of the Keyline classification of land forms. All similar ridges, of no matter what size or shape, are likewise always classed as "main ridges".

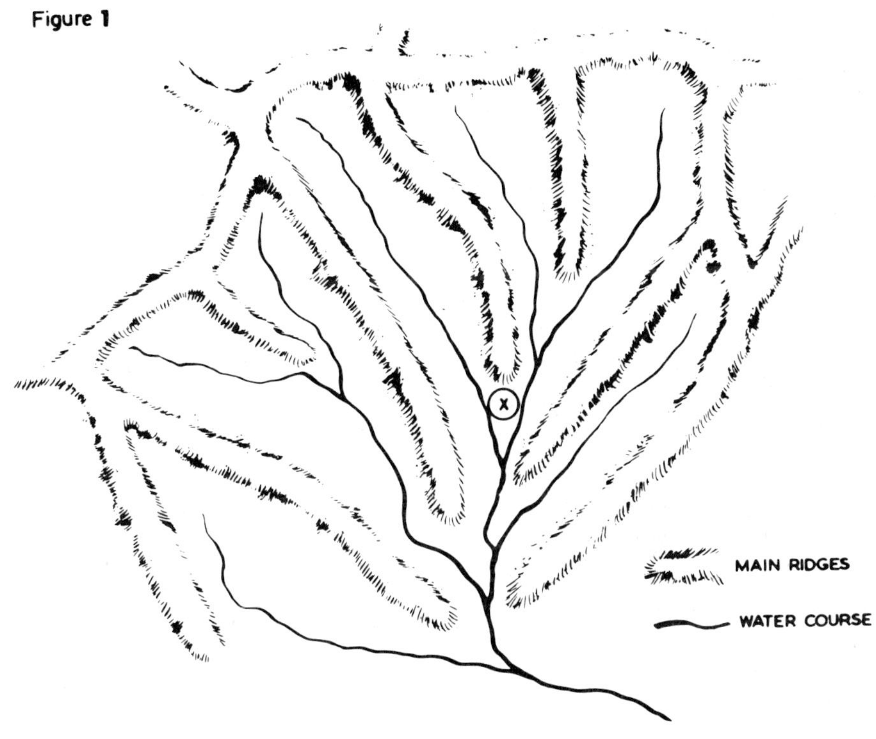

The branching pattern of the 'main ridges' is similar to that of the water courses.

A main ridge has its start between two adjacent streams and at the junction of them. It always joins another main ridge at its upper limits and thus becomes a part with still other main ridges of a continuous water-divide. The connecting patterns of the main ridges are continuous in a manner similar to that of the persistent branching and joining designs of the streams and rivers. So much is that so that if a hollow model of undulating land were turned upside down, the new land form thus depicted would show the stream courses as now the main ridges, and the main ridges as the stream courses, and with their appropriate joining or branching design. However, the main ridge lying between two adjacent watercourses is the dominating land shape in the Keyline classification of the topography of farm and grazing lands.

Just as nearly all land may be said to lie between watercourses, so all the shapes of land lie between watercourses, and are therefore closely related to the main ridge. That holds good for all types of land in all types of country.

The primary valley

Moving in the upstream direction along the main ridge, a succession of valleys separated by smaller ridges falls from the main ridge and on both sides of it to-

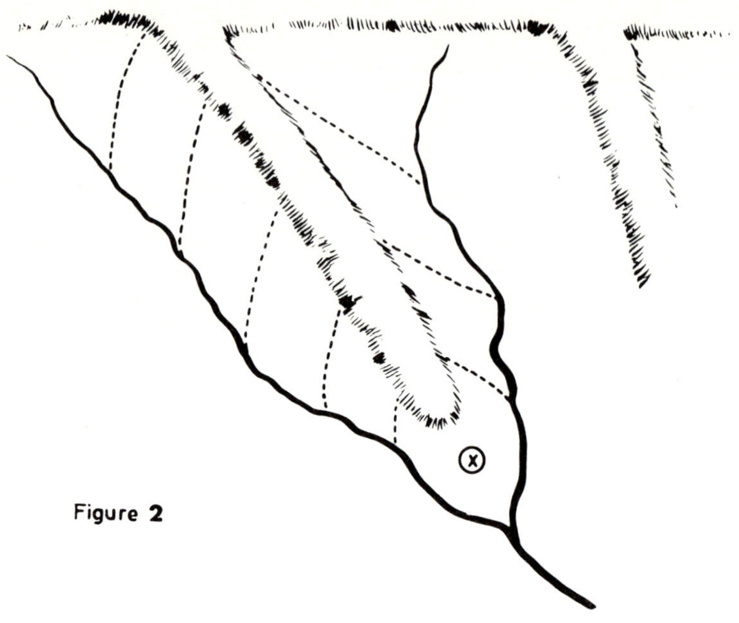

The land between two creeks showing the "main ridge". The dotted lines are the centre lines of the two series of "primary valleys". Between the primary valleys lie the "primary ridges".

ward the appropriate stream course below. Those valleys are the "primary valleys" and the ridges, the "primary ridges". (See figure 2.)

A primary valley head generally starts as a more or less sudden steepening of the side slope of a main ridge. Farther down, the valley changes to a flatter sloping floor which continues more or less uniformly to the stream course below it. The primary valley is the smaller of the valleys of land, and unlike the valleys belonging to creeks and rivers, does not usually have a washed out or channelled watercourse down the middle of it. The rain falling on the main ridge and near each primary valley is concentrated by the valley shape and finally flows down the smooth and rounded valley bottom to the channelled stream course below it.

The primary ridge

On each side of each primary valley lies a primary ridge. The two primary ridges give the primary valley between them its form or its valley shape. On the other hand it would appear that the primary valleys on either side of a primary ridge are the factors which give the form to the primary ridge, since it can often be seen that the primary ridges are the relatively undisturbed portions of the side-slopes of the main ridge, and remaining after the primary valleys have been scooped out, as it were, in the sides of the main ridge.

Referring to figure 2, *x* is the start of the main ridge as discussed. The

120

hatched lines illustrate the branching and joining pattern of the watershed or divide lines and therefore of the main ridges. The unbroken lines are the stream courses and the dotted lines represent the centre line of the primary valleys of the lesser watershed. The primary ridges lie between the dotted lines.

The secondary valley

On occasions a series of primary valleys on the one side of a main ridge will join up with a larger valley which does not contain a channelled watercourse in the bottom of it. Such is named a "secondary" valley, and it will have at its start its own Keypoint and Keyline.

The land which may lie between two such streams as in figure 1 (page 48) may vary widely in area. Rainfall and the degree of slopes of undulating country have a very considerable effect, and generally the area will be smaller in the wetter, and larger in the dryer zones.

For those reasons one or more main ridges or parts of them with their primary valleys and primary ridges may be present on the one property, while another of equal size may include only a small part of a main ridge and one or two primary valleys and primary ridges; or again a property may contain only a portion of one of such lesser land forms. However, because the size of those land units usually relates to the annual rainfall, with the smaller units in the wetter areas and larger in the drier areas, and since the general size of farm and grazing holding rises as the rainfall decreases, the general order is that several of all the forms exist on each property. There is thus a need to devise planning relationships between the various grouping of those land forms.

Classifications of land forms are according to their shape and not to their areas. Thus it will be seen that yet other pertinent descriptions of land may be defined. Land can be said to have "length", which would be from the water-divide on the main ridge to the stream course below. Such length-of-land may be as "short" as 402m, or "medium" as 0.8km to 1.6km, or "long". "Long" or long-slope country may range up to some kilometres from the top of the main ridge to the stream course below it. Coupled with the length of land, slope of land may be designated. The land-slope again relates to the main ridge and to the stream course below it.

LOCATION OF KEYLINES

Reference has been made already to the two slopes of a primary valley: from the higher slope of the main ridge, the primary valley forms with a first slope which may be short in relation to the full length of the primary valley. The first slope of a primary valley is steeper in relation to the second longer and flatter valley slope. I named the point of change in the two slopes of the primary valley, the Keypoint of the valley, and a contour line from that point and both directions in the valley, the Keyline of the valley. The name Keyline, coined to describe that contour line, has also been given to the full system of land development.

A primary valley, bounded by the portion of the water divide of the main ridge above it and by the water-divides of the primary ridges on either side of it, is the primary, the smallest, or the first catchment area. Likewise the valley or the drainage area of a small creek would be bounded by the water-divides on the main ridges on either side of it, and by the portion of the upper main ridge which is joined by those two. Those distinctive drainage areas progress in size until they eventually embrace the complete watershed of the largest river systems to which all the waters flow.

Now, while it is customary to classify, plan and develop great tracts of land for government purposes according to the drainage areas of the larger river systems, that could not apply to smaller developments. The subdivision, the planning and the development of land for the individual farming and grazing holding should not be according to their lesser drainage areas. For instance, in the development of farm water resources it is often a great advantage to inter-connect the smaller drainage basins by diverting runoff rainfall from one such drainage area for storage in another. Similarly, in the working of land with cultivating and harvesting machinery, water-divides are often crossed over. Therefore the more suitable boundaries for both general working and for maximum development are those of the watercourses themselves.

The rising relationship of Keylines

The above fact may again be illustrated by considering the water position in two separate areas. In the first instance, all or several primary valleys falling from the one main ridge to the creek below may be contained on a property where, while the rainfall is in the order of 508mm to 1016mm a year, there exists at frequent times both intermittent water shortage and later on considerable rain-fall runoff.

Primary valleys on undulating land contain the suitable sites for water storage, but those sites may not have satisfactory catchment areas within the catchment of their valleys. Water may be then diverted from one or from several primary valleys to fill a dam constructed to receive the water. The highest suitable site for a storage dam in a primary valley of good shape for the purpose is just below the Keyline of the valley, where the wall for the dam would cause the water line of the dam to coincide with the Keyline of the valley. Also a series of primary valleys falling from the one side of a main ridge will have a rising relationship one to the other as the main ridge rises from its start at the junction of two watercourses. The Keylines of the series of primary valley will also have the same rising relationship. Therefore, where some of the valleys are of satisfactory shape for water storage construction, it can be seen that the whole series could be interconnected by channels diverting runoff from valleys with no dam to the storages. Also, the overflow of the higher dams can be diverted to one or more lower down in the chain-like series. (See aerial picture of Yobarnie dams and channels, *Plate 5*). The Keylines of the primary valleys may not, on the other hand, be suitable as dam sites, but the same type of inter-connected series of dams may be suitably placed at a somewhat lower lever. So long as the channel connecting the series of dams has a gradient which follows the fall of the land and at a lesser slope than that of the stream course, the succession of dams will generally have progressively more land below them which may be suitably placed for the lower costing gravity-flow irrigation systems. As a general rule, the direction of the flow of diverted water on farms and grazing properties should follow the general fall of the land. The direction is clearly shown by the fall of the watercourse below.

In a second instance of water diversion from the one catchment to and through other catchments on a farm, considerable water from runoff rainfall may be entering from land above via a small creek. There may not be a good dam site near the entrance, or there may be more water constantly available than a storage near the creek's entrance to the land will hold. That water may then be diverted through a whole series of primary valleys for filling storages constructed

122

only in those valleys where there are suitable sites for dams. An example on those lines is also illustrated on the forementioned aerial photography of Yobarnie (*see Plate 5*). The entrance of the small creek with its relatively small diversion dam is to be seen on the western boundary. The diversion channel from that dam fills in succession four more dams. Note the increasing distance of each of the succession of dams inland from the creek which feeds them, and note also the increasing areas below each of the dams and between them and the creek.

Contour maps for land planning

These shapes of land and their water relationships are considered to be of paramount importance as basic elements of land planning. From Chapter 2 of The Geographical Basis of Keyline by J. Macdonald Holmes, I quote:

> Not only must land scientists and farmers have an eye for country, they must also "see" their land on paper, since working on land and planning on paper must be performed together systematically. The best way to show land shape on paper is by means of a contoured and scaled map. Contours on paper represent horizontal lines on the ground, or better still, outcrops of horizontal surfaces on a slope. They also denote heights above some fixed base level, usually sea level. Contours marked out on the ground would be level lines. On paper they would twist and turn, depicting valleys and ridges, steep land and level land.
>
> One can appreciate the shape of land from a suitable contour map of the land much better and more readily than one could from an examination of the land itself. Although the lines appear to be indiscriminately spread, this is not so. They follow very definite natural patterns. Within the pattern of these contours lie some of the secrets of Keyline . . .

As Professor Macdonald Holmes has said — the best way to illustrate land on paper is by means of a contoured and scaled map. The "contour interval" which is the distance apart of the contours vertically, should be suitably selected to disclose the shape and the form of the particular land, and the scale should be adequate for paper planning. Indeed, with the aid of such a map of an individual property, an experienced Keyline planner could lay out a working specification covering the full development of the property, provided that he had a good knowledge of the climate and also had the landowner available to give details of the soil and other materials which the planner would need to know about particular locations on the land itself. Such a plan could then be produced in ample detail to include the manner and type of the water resources development, as well as the location and the most suitable size and wall heights of water storage structures the sites and sizes of water diversion channels and the location in relation to such works of irrigation areas. Even the appropriate irrigation procedures could be designated.

Stock watering points, farm roads and the positions to leave trees already on the place or to plant them, the sites for farm buildings, for subdivision fences and stock working paddocks — all could be decided quite competently without the architect of land development actually seeing the land itself. The plan could then be transposed onto the land either as a project or piece by piece as circumstances may permit or dictate. Such are the value of good contour maps!

THE GEOMETRY OF KEYLINE

We may now proceed to further illustrate, by means of these contour maps, the various land shapes and relationships already described, and also by the same means to show the principles of the intimate relationship of the contours of the primary land forms and the uses of them which are disclosed in the Geometry of Keyline.

Figure 3 (page 55) shows a portion of a main ridge with two primary valleys and primary ridges on one side of it, falling from the main ridge to the watercourse below. The general fall of the land is seen in the direction of the fall of

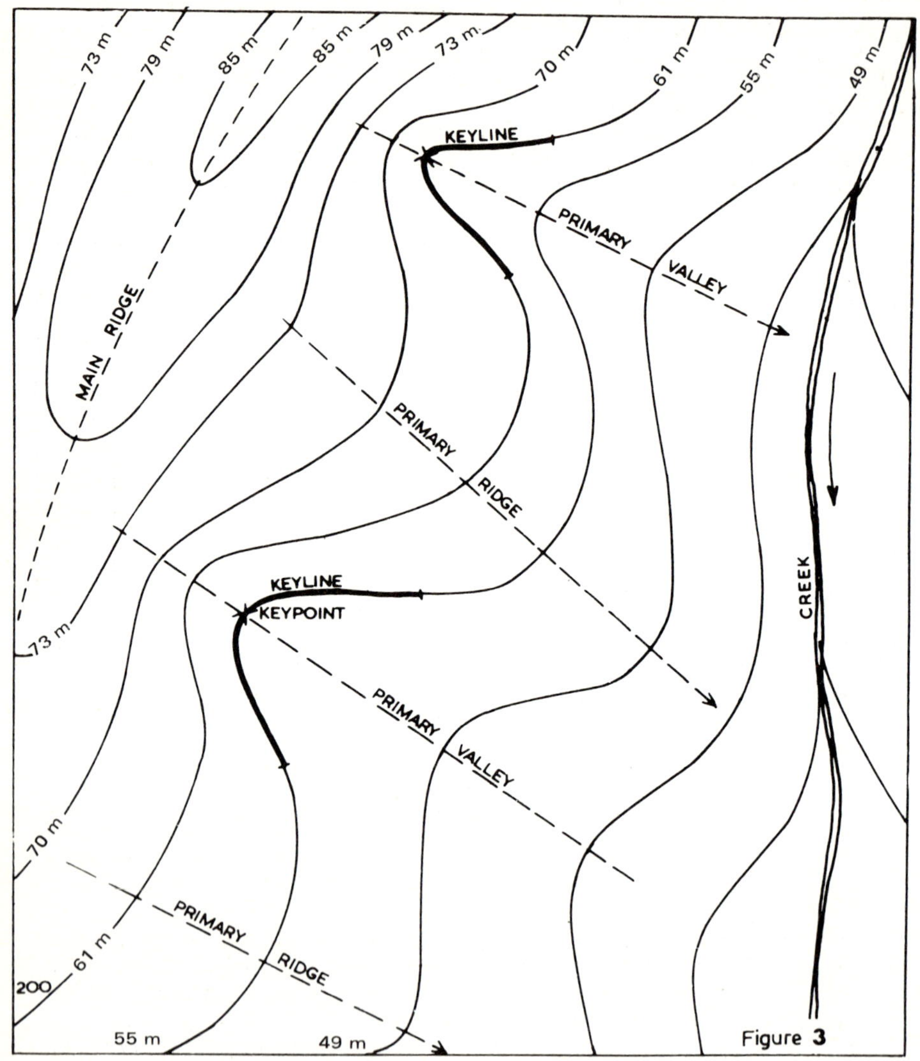

A contour map showing two primary valleys falling from the main ridge to the watercourse below and with a primary ridge between them.

the creek. The slope of the land, the cross fall from the main ridge down the primary ridges to the creek below, is indicated by the arrows on the primary ridges. That is also the "length" of the land. The Keylines of the various primary valleys are marked in to illustrate the rising relationship of them as discussed earlier.

Figure 4 (page 56) shows the more intimate contour relationship between a primary valley and a primary ridge. In the somewhat normalised contour diagram the contour lines of the primary valley show, by being closer together at the top of the map, the first steeper upper slope of the valley, and the Keyline of the valley marked by a thickening of the appropriate contour line. Below the Keyline the wider interval between the contours indicates the flatter valley floor below the Keyline of the valley.

In the primary ridge portion of that contour diagram the distance apart of the contours illustrates that the centre of the ridges is flatter than the sides of the ridge out towards the valley. It is the typical shape of a primary ridge.

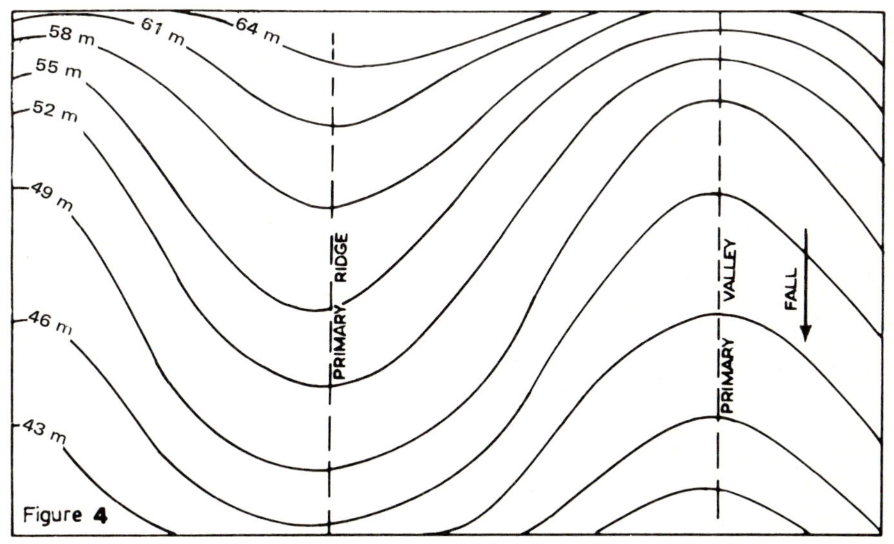

A contour diagram showing a primary valley and a primary ridge.

Flowing water makes its own pattern

The contour diagram of figure 5 is repeated in figure 5 (page 58) to illustrate the pattern of water flowing over land. Water in very heavy rainfall will be moving over practically all the land surface at the one time. The upper of the two diagrams of figure 5 illustrates the pattern of water flowing in relation to the contour shape of the land. Water follows the steepest path downward to the valley below it, therefore it flows at right angles to the contours, as that is the shortest distance between contours, and the steepest path. The water in effect continuously adjusts its downward path because of that fact, and so results in

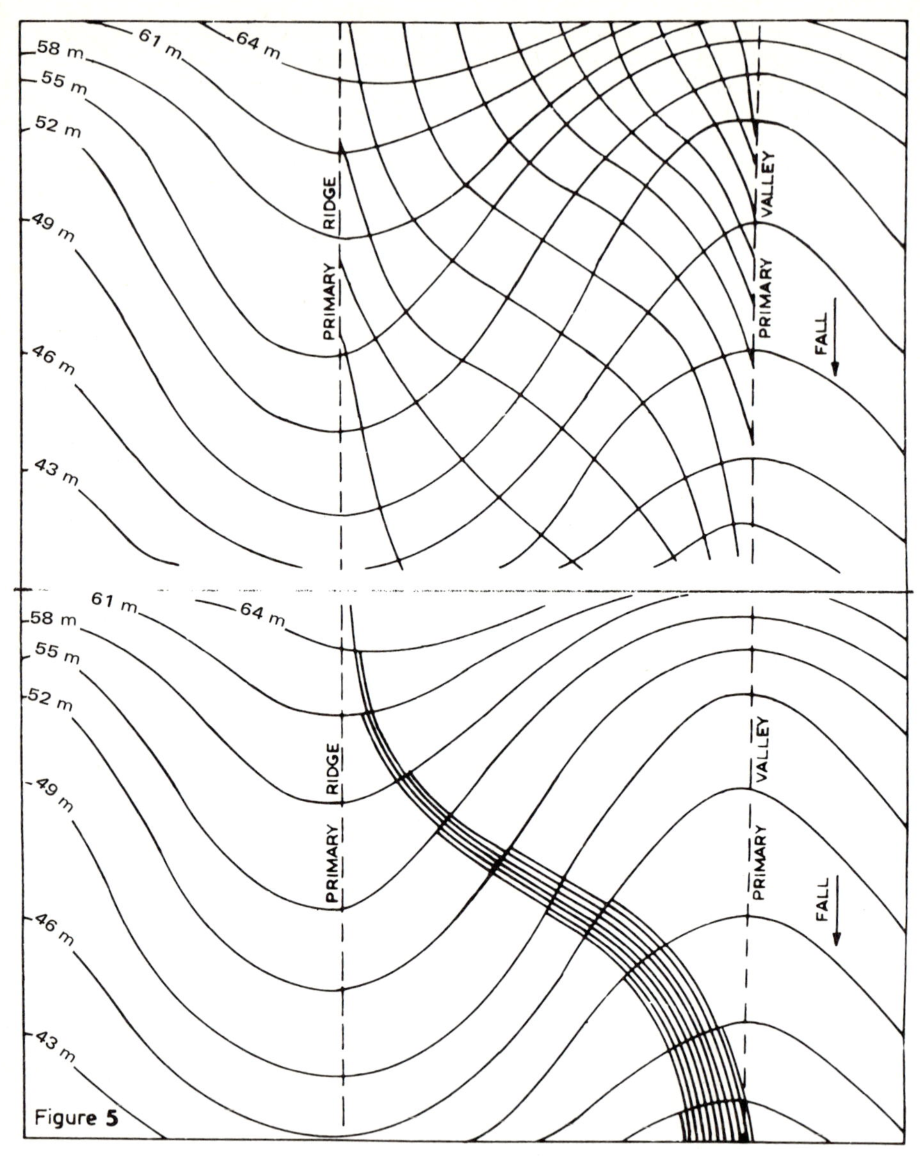

The contour diagram illustrates flowing water.

Upper: *The flow paths of runoff water from the ridges to the valleys are flat S curves.*

Lower: *Depicting one flow path to illustrate the increasing volume of flow from the ridge to the valley.*

a general flat S bend form to the flow directions. The lower diagram of figure 5 illustrates the factor of increasing flow from ridge to valley.

The natural results arising from the movement of water is that from the time the rainfall produces general runoff the land is not "watered" evenly, as more water progressively is in contact with the land longer in moving from the ridge shape to the valley shape. One effect to be seen in the dryer weather is the greener valleys and the dryer ridges.

KEYLINE PATTERN OF CULTIVATION

The Keyline pattern of cultivation was devised to use more effectively and evenly over all the land shapes the first of the flowing water which results from rainfall which produces runoff. The same technique also provides the means for evenly spreading the water in the system of "hillside" irrigation which I devised and named Keyline pattern irrigation. The technique of pattern cultivation or pattern ploughing takes a selected contour as a guide and proceeds by working either up the slope or down the slope, according to circumstances, parallel back and forth from that contour. Because contours are not spaced evenly apart and are not parallel on the surface of the land, any cultivation which works parallel to a given contour line marked in on the land surface must inevitably drift off the true contours as the cultivation continues in that form. That fact is used as a device in "Keyline cultivation" to break and reverse toward the ridge the natural flow of runoff rainfall in whichever direction the farmer desires. As the flow from rainfall runoff increases, it will eventually reach proportions which overcome the drift of the water which has been induced by the pattern of cultivation, and so resume its natural flow path as illustrated in figure 5 (opposite).

A contour diagram illustrating Keyline cultivation for a primary valley. The broken lines show cultivation parallel to the Keyline both up and down the slope.

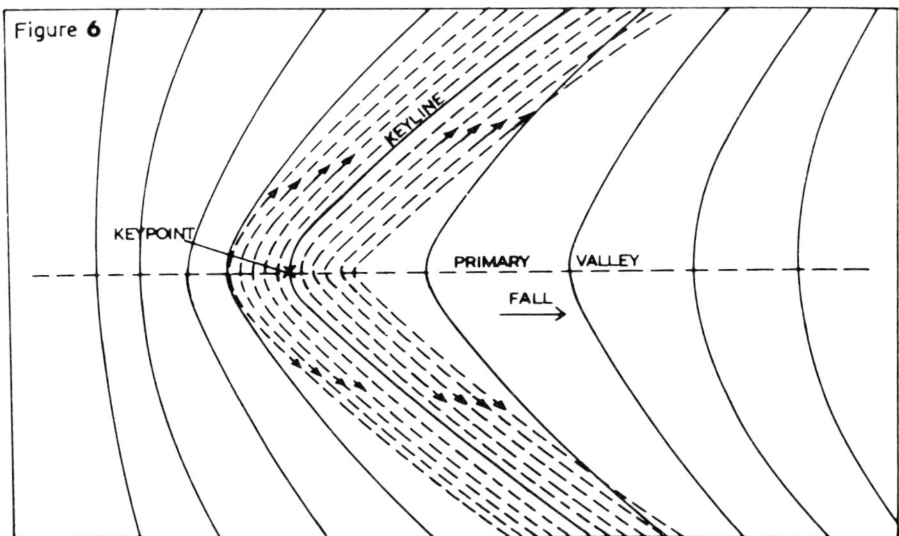

Keyline cultivation has practical and positive application on any of the shapes or parts of the basic shapes of land wherever it is desirable to spread water uniformly, or to cause shallow flowing water to drift one way or the other.

Since the non-parallel form of contours when marked on the surface of the land permits the use of the technique of Keyline cultivation, it follows that the only circumstances where the selection of the correct direction upward or downward of such parallel cultivation is not of significance is where the contours occasionally form perfectly parallel lines on the land surface. In those rarer circumstances, cultivation done parallel to the selected marked contours will have the same water-spreading effect, whether they are done parallel upward or parallel downward to the contour lines. Keyline cultivation, or pattern ploughing, is illustrated in its application on various land forms in figures 6, 7, 8 and 9.

Figure 6 (page 60) is a contour illustration of a primary valley with the Keypoint of the valley and the Keyline of the valley marked in. The broken parallel lines illustrate the correct procedure for cultivating the valley area. For the portion of the valley above the Keyline the cultivation parallels the Keyline up the slope of the land. Because the cultivation reaches the next contour above the Keyline first as a point in the centre of the valley, it follows that the general slope of all the small furrows is downward away from the centre of the valley.

The first and steeper slope of a primary valley is usually much shorter than the main flatter slope of the valley below the Keyline. In some circumstances the upper and steeper slope may be too steep for the working of cultivation implements. Wherever the slope can be worked the ploughing can be extended sideways well out and across the primary ridges on either side of the valley and continue to have its water spreading-to-the-ridge-effect.

Below the Keyline of the valley cultivation always proceeds in the opposite direction — by parallel workings downward from the Keyline as illustrated. Here, because the cultivation parallels downward from the Keyline and reaches the next lower contour first at points out on the valley sides, it again follows that the general slopes of all the small furrows are downward away from the centre of the valley. The cultivation of the primary valley below the Keyline is best not continued downward too far. Another contour line can be selected for a cultivation guide lower down the valley, for instance the 40m (130 ft) contour of the diagram, and the cultivation then proceeds downward from the new contour guideline. The area in the centre of the valley above the line, which is not cultivated when the work from the Keyline downward reaches the guideline on the sides of the valley, can be ploughed out without any worry about its pattern.

The Keyline cultivation of a primary valley below the Keyline cannot always be extended as far out on the ridge as can that above the Keyline, as the general pattern of primary ridge cultivation is parallel upwards to the selected contours. By extending the cultivation of the valley pattern, for instance to the centre of the primary ridges, the result could be that the water would first drift out of the valley toward the steeper sides of the primary ridge, and beyond that point the furrows may tend to drift water back from near the centre of the primary ridge toward the steeper parts. Thus water could be concentrated adversely from both directions towards the steeper areas.

It can be seen from that there is an area where the contours indicate a boundary between the primary ridge and the primary valley shape, and which lies down the steeper sides of the primary ridges. And so any combined cultivation of those shapes must be adjusted accordingly.

Some primary valleys cannot be cultivated as described because the valley contour shape may be such that it becomes impossible to make the turns in the valley floor with the cultivation equipment. The contours in the valley bottom are too sharp or somewhat pointed. The valleys are most suitably worked on a herring-bone design and with tractor-attached instead of trailing implements. The valley is then cultivated in two parts, with the centre of the valley as the dividing line. Cultivation proceeds from the middle of the valley along the ap-

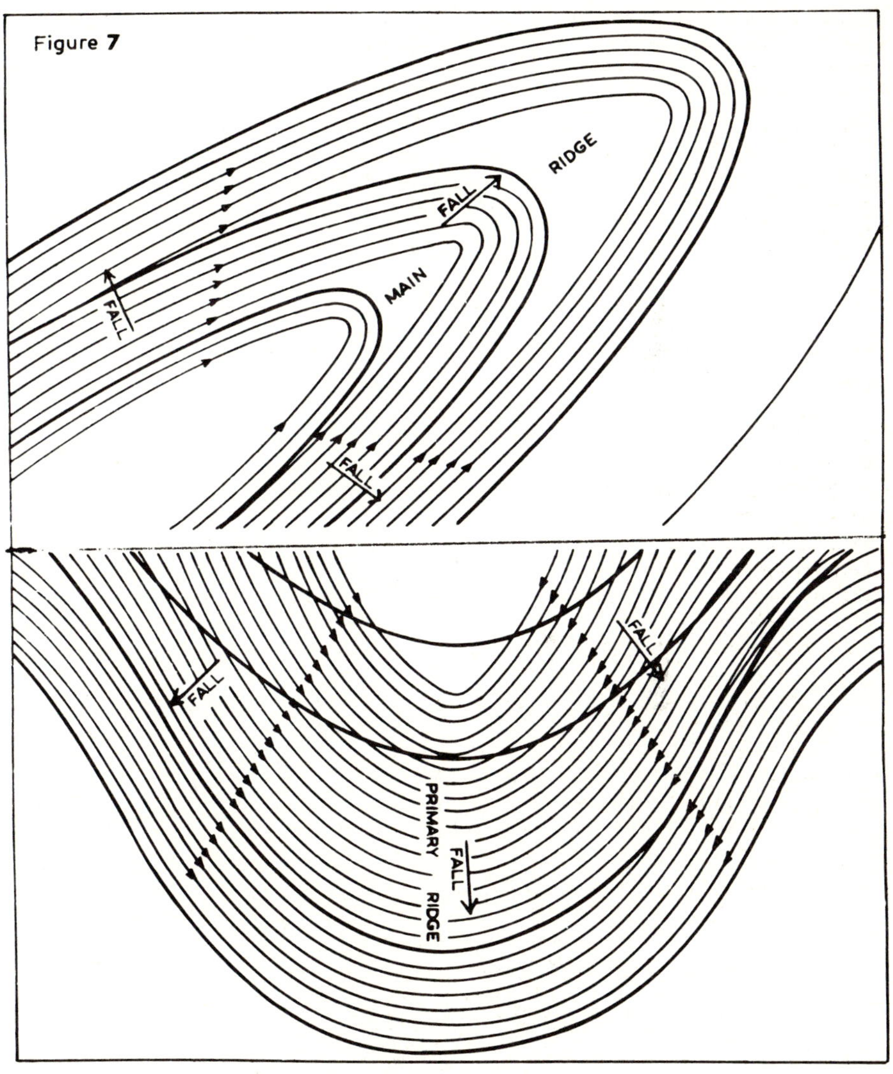

The contour diagrams show a primary ridge and a main ridge shape with Keyline cultivation designed to drift the first flow of rainfall runoff towards the centre of each ridge.

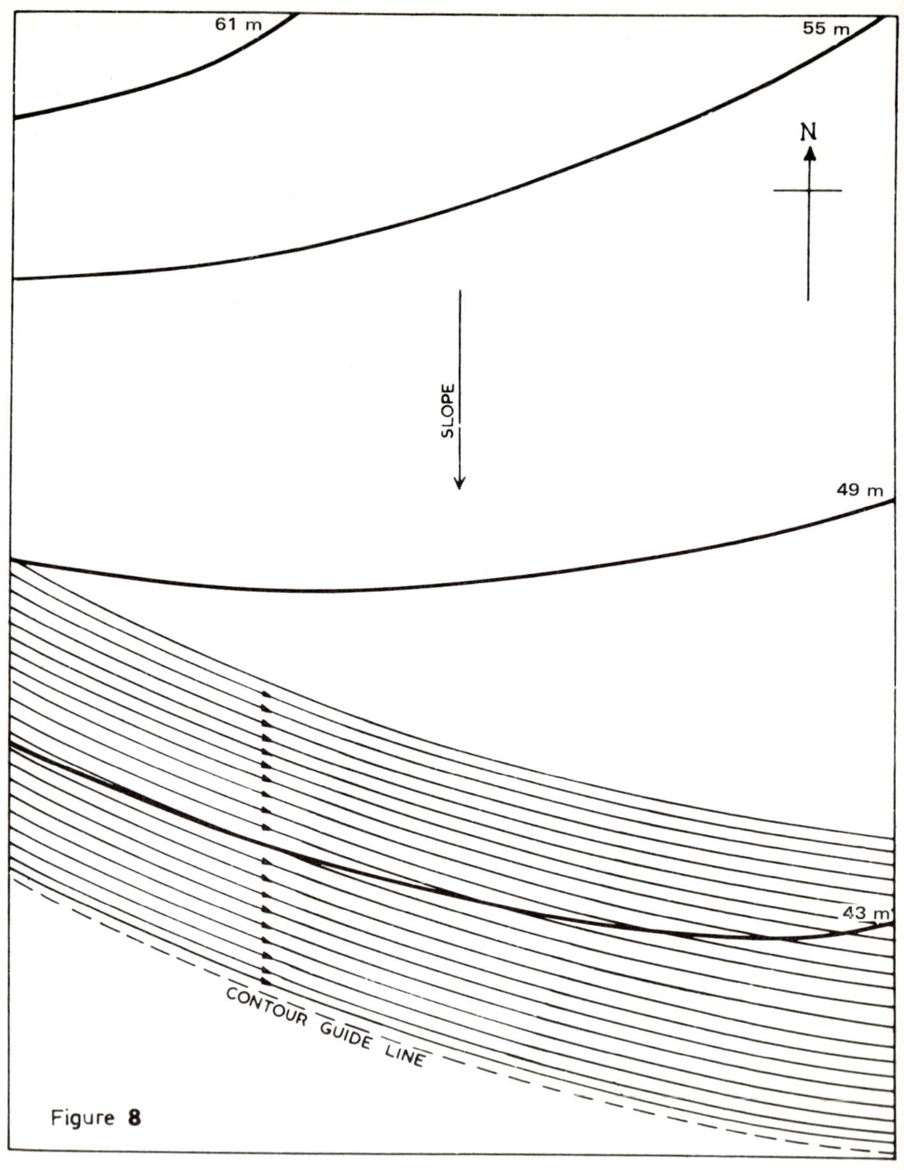

The contour diagram shows a selection of the contour guide line for the Keyline pattern of cultivation which will drift water toward the flatter eastern side of this area.

The contour diagram shows the selection of the contour guide line — the 55 m contour — for the pattern of cultivation which will drift water toward the steeper western side of this area.

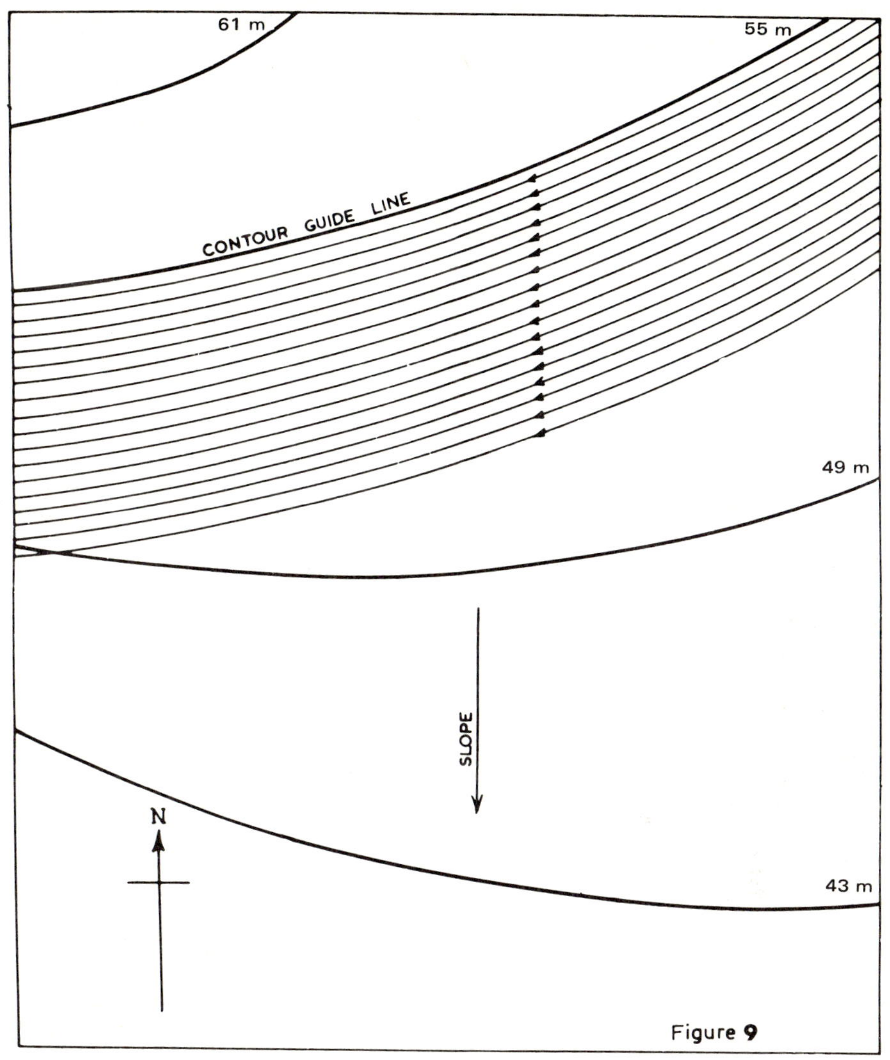

Figure 9

propriate contour and working parallel downward as before, and completing the cultivation first one side of the valley and then the other.

The lower part of figure 7 (page 61) shows the contours of a primary ridge. The natural water runoff from the ridge shape is essentially sideways to the primary valleys on either hand. The effect of Keyline cultivation is to drift the first runoff water towards the central and flatter portion of the ridge. Cultivation which parallels upwards from any contour line achieves that object. The broken parallel lines of figure 7 illustrate a cultivation which works parallel upwards to the contour guideline. As with the cultivation of the primary valley shape it is better not to continue the parallel cultivation too far from the one

131

line. Two or sometimes three contours at selected distances apart should be used to guide the cultivation.

The contour shape of a main ridge above the headings of the primary valleys which lie on either side is essentially the same as the primary ridge shape, except that it is much longer and narrower. The same parallel upwards-from-the-contour cultivation is employed for the main ridge as for the primary ridge in the upper part of figure 7 (page 61).

Figures 8 and 9 are contour diagrams of the same area of land. Together they illustrate the type of selective control of the first flow from runoff rainfall, or for spreading irrigation water, which Keyline cultivation places in the hand of the cultivator of land. Both contour diagrams illustrate a paddocked area on which the slope near the western fence is about twice as steep as the slope on the eastern side. In figure 8 is illustrated the selection of the correct contour guideline and the parallel cultivation appropriate for spreading flowing water from the steeper western to the flatter eastern side. The selected contour is located from the higher of the two corners on the southern boundary fence. Again the whole area need not be cut out from the one line, a second contour guideline being selected halfway up the land, for instance the 49m (160 ft) contour of the diagram. The area below that contour line remaining unworked when the parallel cultivation reaches the 49m contour on the western end is cultivated out by continuing upwards or by any convenient means. The effect, even if opposite to that desired, will have little issue against the general pattern.

Figure 9, on the other hand, illustrates the selection of the guideline contour and the parallel downwards form of the cultivation, if it were desired to drift any flowing water towards the steeper land on the western side of the paddock. The contour guideline is then located from the lower of the two corners on the northern fence, and cultivation parallels the line downward.

In actual practice Keyline cultivation would not be used for that purpose as it would concentrate water on the steeper area. It is therefore given only to further illustrate the degree of selective control which a knowledge of the geometry of Keyline places at the will of the landsman for the betterment of his own soil.

CHAPTER SIX

Water categories of Keyline

WATER PROVIDES THE MAJOR PART of the value of all agricultural land, and probably as high as 95 percent of it in very many cases. In the drier western country of New South Wales, land with only 254mm to 305mm (10 to 12 ins) of average annual rainfall as its sole water resource may be worth little more than $2.50 to $5.00 an acre. The same type of land or indeed land less rich minerally, but enjoying an annual rainfall of 660mm to 762mm (26 to 30 ins), could be worth $30 to $60 or more an acre, or 20 times as much, if a buyer can be found in today's difficult economic circumstances. It follows therefore that to allow water to run off a property needlessly is waste of a high order.

When water is stored on a farm for later use it can become, with the aid of good design, its own great work force, because as it moves down through the soil it draws air in behind it and then, because of its flow volume, it distributes that first force in irrigating the soil.

Too little water in the soil restricts and stops plant growth. But too much water in the soil over long periods has a similar destructive effect. In the soil there must be movement of water so air can return to it. It is necessary so as to enable the soil aerobes to continue to process food for the roots' cells to feed the plant.

A pasture overwet for too long may wilt, as it does through lack of moisture, and soon its composition is changed and it is no longer good stock feed. It may now be good insect food and then pasture pests invade it. And if waterlogging continues the pasture grasses will die to be replaced by water weeds which get their air through hollow stems.

In the too-slow movement of water over the land, growth of valuable plants is restricted and lost. But if water moves too fast and in too concentrated a stream it becomes an earthmoving force which will eventually wash away the soil.

Water movements in soil

Again, there are two distinct movements or rates of movement of water into soil. When water is applied rapidly in irrigating a soil in need of water, there is first, a very rapid penetration of the aerated topsoil. Immediately on the saturation of that soil, which may be only a few centimetres deep, there is second, a very sharp decline in the rate of water penetration, as now more water cannot enter the top soil faster than it moves from the surface layer into the soil or subsoil below. The second rate may be more than 100 times slower than the first. When that occurs it is better that excess water flows off the land altogether, than the soil be allowed to become oversaturated.

Water going down through the soil in excess of the soil's requirements may overcome the soil processes which hold on to the finer minerals against the fertility robbing effects of leaching.

The proper planning for the full development of land should be preceded by an evaluation of the water resources available for the development. For the purposes of such evaluation, farm waters may be divided into four categories and the various relationships within the categories considered.

The first category of water

Water of the first category which belongs to the land is the rain which falls directly on it and which is absorbed by the soil. That water is of the very best quality and its price is the lowest. Moreover it is the farmer's own; no outsider can rob him of it and no one can turn off the tap. Only the farmer himself may rob his soil and his land of that high quality and free water, by using methods of soil and pasture management which prevent its maximum beneficial use.

The second category of water

Water of the second category relates closely to the first category. It is the water from direct rainfall which has filled the soil for the time being and now flows over and off the land. In conditions of general rainfall runoff the water moves from the ridge and valley areas of the farm to flow via the valley bottoms to the creeks and is so lost. By and large the runoff water from the farm is also the farmer's to do with as he wishes. He may control and store it and use it or allow it to run to waste. It is usually of good useful quality for all purposes.

The third category of water

Water of the third category originates outside the farm or grazing property boundary, and flows onto the farm usually via the primary valleys and the stream courses. It may leave the farm without even being seen. That water on the farm is not entirely the farmer's own, as others with some right to it may be interested in using it. Before that water can be used or stored for use by the landholder it is normally necessary to obtain a licence from a government authority. Apart from preserving the rights of other possible interested parties, the authorities are committed to assist in the improved use of water and generally act towards the farmer in a helpful capacity. The water is usually of suitable purity for agricultural purposes, including irrigation and household use.

The fourth category of water

Water of the fourth category is ground water, which comes from beneath the surface of the land itself. It occurs in springs, wells or sub-artesian or artesian bores.

There is no other water.

All agricultural holdings possess water of significant value in some of those categories, and there is no other source of supply.

For the purpose of subsequent pages of this book those water categories require additional description. Water of the first category, that is direct rainfall which penetrates into the soil, can be the only, or the principal, source of water available in some areas. There may be no runoff from rainfall, no flowin from outside, and no ground water. On occasions the pasture improvement of some land of previous runoff record may so increase the entry into the soil of a meagre rainfall as to prevent runoff, and make it necessary to leave some areas undeveloped so they will provide runoff water for filling stock dams. In other rarer, but similar circumstances, it may be necessary to surface treat and seal some areas of land to turn them into artificial catchments to provide for stock watering and for homestead use.

Methods of soil improvement designed specifically to take more water into the soil would appear to be useless in the above conditions. However, since the general system of grazing management tends inevitably to reduce the adequate aeration of the soil, cultivation of the right kind will improve the methods by improving soil aeration and result in better quality pasture and more of it, without extra water coming into the picture at all. Moreover, improved management of pasture on its own can likewise improve the soil and both the production and the quality of production, still without more water being made available. Again any result from planning or design which retards the drying action of wind is another improvement.

However, as rainfall generally does produce runoff the first consideration in the improved use of direct rainfall may be from soil treatments which cause more of it to enter the soil. A three-year course of soil and pasture improvement is a suggested program to follow. However there is no point, and only additional expense and loss of water, in opening the ground so deeply it is forced to take in more water from direct rainfall than the soil can effectively make use of at its present stage of development. Deep soil, which is also fertile, can take in and effectively use much more water than can a shallow soil which is only in its first year of treatment.

From this discussion it is immediately apparent that low-costing cultivation may give improved use in the soil of direct rainfall, but it also follows that because of the manner of soil improvement, the amount of rainfall runoff will be reduced, and on occasions even drastically so.

Water of the second category, namely rainfall runoff from the particular farm itself is seen, for the immediately foregoing reasons, as liable to be affected by soil treatments which may induce more water from direct rainfall to enter the soil; and in some circumstances that may have an effect on the planning of farm development.

Water for storage and later use

The water from rainfall runoff on the farm is water which cannot be used at the time it first becomes available. To make use of it the water would need to be diverted to one or more storages built for the purpose, and be held there for later use in irrigating the land when the soil has need of it. At this juncture it would seem that the critical planning details of locating the diversion channels and the sites for storage dams for the water would need to be immediately considered. But that is not so as no artificial water lines – channels – for water control and use should be fixed until all the sources of water are fully considered.

Water of the third category flows onto the specific area from an outside source. There are two general modes of its arrival.

First, the source which is likely to provide the most water is a constantly flowing or intermittently flowing stream which receives its water from a catchment situated farther up the land outside the farm.

Second, primary valleys may be partly on and partly outside the farm, and rainfall runoff from the catchment area outside flows to the farm via one or more of the divided primary valleys. In that event the Keylines of those primary valleys are also outside the farm.

Water of the third category is now seen to be affected by the farm rainfall runoff water of the second category, which joins up with it on the farm itself. Water flowing to the farm via stream courses is normally joined by farm runoff water flowing to those streams via the primary valleys of the farm and, if not prevented from doing so, both waters flow together from the farm to be lost.

Water of the fourth category – ground water – may be made available to the land from just below the surface or be brought up from hundreds of metres down. While gound water generally may play a less important agricultural role now in the better rainfall districts, the incidence of artesian and sub-artesian waters, below vast areas of land of only a few millimetres of unreliable annual rainfall, has permitted carrying great numbers of sheep and cattle in conditions otherwise impossible. But usuable and very valuable ground water does exist also in many areas of better rainfall where the unpredictable incidence of dry spells and of shocking losses from severe droughts, take a serious toll on constant production. It appears that, except for the fairly wide official knowledge of the artesian and sub-artesian basins, ground water generally may be an enormous untapped and undiscovered agricultural resource. Its discovery and investigation must be considered as generally outside the capacities of the lone farmer, even though ground water could exist below his own land and of a quality and quantity permitting its profitable use.

In both quantity and quality ground water has wide variations. It may be available only as a mere trickle or up to rates of flow of a hundred thousand gallons an hour and much more. It may be so salts laden as to be useless for any purpose, or vary from that condition through to a quality suitable for watering a vegetable garden. Much ground water of good quality is associated with the gravel and sand beds of present day and geologically ancient river systems and old lake sites. In some cases the past geological formations showing possible sources are disguised by the present surface solid geology.

Other than for stock watering and household uses ground water in some cases can have substantial potential for irrigation. First, the water may be available in sufficient quantity for direct pumping from wells or bores and be applied to the land by the type of irrigation best suited to the land form. Second, the water at its available rate could be accumulated in a storage dam constructed for the purpose, where continuous day and night capacity pumping would provide the maximum quantity of water. The water stored from 10 to 12 days pumping may then be spread over an irrigation section of land at as fast a rate of flow as the land form would warrant.

These foregoing descriptions of the Keyline classification of the water of farm and grazing land, while not appearing in this form in my earlier Keyline books, were presented by the author in a paper to the Sydney conference of the Australian and New Zealand Association for the Advancement of Science 20-odd years ago.

The assessment of the waters available from categories 2 and 3, which may usually provide the principal irrigation potential, appears at first glance to present considerable difficulties. Indeed, that is so, as the class of information which may be available from government departmental sources has never envisaged the maximum development of that vast collective water resource. In the eastern States of Australia, information is sometimes available from departments on the minimum annual runoff. That information is considered to be of value to an owner of land who may wish to construct dams of assured filling capacity for stock watering and/or irrigation.

But water for those two distinctive purposes involves completely different and even opposing considerations, and to such an extent that information of value in planning stock watering requirements may have no bearing whatever for the planning of farm irrigation developments. For instance, a first gauge to the value of stock water storages is that the water never runs short; each storage must be relied on over the extended periods of drought. If the same considerations were applied as the basis of design against each water storage of a projected farm control and to use as much as possible and profitable of the water which became available, then one of two things would become a certainty. Either the full project would be abandoned before it started or the planned development would envisage the waste of most of the water.

THE TWO COSTS OF WATER

On my own property, complete "reliability" would have cost as absolute water waste, at least 70 percent and probably 85 percent of the waters from categories 2 and 3.

Every kind of water resources development has two major costs: The cost in money and the cost in water itself. And the latter cost always reaches the greatest water waste with the highest degree of reliability. Complete reliability means that no matter what the weather conditions or how severe and long the drought, there will still be water remaining for all its designed purposes.

That type of reliability is appropriate in the storages for the water supply of a large industrial city. It does not matter if four-fifths of the stored water is still available in the dam, and therefore wasted when rains fill and overflow the storage, and again may waste much more water in the second manner of overflow than the full capacity of the storage. In a similar fashion the large government irrigation projects with their vast storages and equally costly channels for water distribution and control are designed to have a high degree of reliability and therefore to waste enormous quantities of water. It could be argued that must be so as large towns which were made possible by the water depend absolutely on the reliability of the irrigation water, and the only matter for consideration is that sufficient water is always available to fulfil its designed purposes. It is physically impossible not to waste water, so does waste of water matter to anyone?

Planned waste of water, where it is necessary to have "reliability" of water supply for the large city, is an essential and justifiable basis of design, and perhaps it is so also, if to a lesser extent, for the large government irrigation area projects. But it cannot be a sound basis of design for the development of farm water resources where they involve irrigation. And particularly is that so in those circumstances where, even with all available water controlled, stored and used, the limiting factor in production would still be lack of water. Is not that the position on so much of the agricultural land in this country which does have undeveloped water resources?

The attitude of mind of the farmers and graziers to the water resources of their own land is of paramount importance to the nation-wide development of these resources. The traditional ideas about water from the older countries with climates less harsh than our own, and the many academic and scientific dogmas and conceits relating to water generally, should now be abandoned in favor of some realisitic and practical thinking and acting on these matters.

For instance, an approach which demands complete reliability of farm irrigation water supply will often limit the area to about 6ha of high cost per hectare irrigation, in circumstances where a like amount of money would start with a much larger area of irrigation which could be progressively expanded. It may not only cost considerably less a hectare to produce but so much less a hectare to run as to be profitable for the production of a greatly increased range of products.

Stored water, a second bank account

In the development of his land's own water resources why shouldn't a farmer be able to think of his stored water as a second bank account? Provided he can take the water out of the dam quickly and at a very low cost and can spread it over his land with similar speed and low cost, he can thus "trade" the necessary water for the crop he wants or the pasture he needs for his stock. And why can't he continue to do just that and, if need be, use up all the water of the storage? True, he is then without the water, but he has more than the value of the water in what the water had "bought" in crops and pasture. An empty water storage dam then is not the sign of failure but quite the reverse. Rather would a filled dam in a drought be a sign of failure. If the owner could not then get the water cheaply on to his land and save his stock because it would now be too expensive, the full dam is not "success" but a badge of failure.

However, the position of farm-developed water resources, which involve the irrigation of a substantial area of the holding, would rarely involve such a use-or-hold decision as this instance would indicate. In the smaller or larger complex of storages and their respective irrigation paddocks, which would almost invariably be involved in a farm water development for irrigation, some dams will be larger than the others, one or more may have a lesser area to serve in relation to its storage capacity, while another may have very favorable replenishment features. Therefore in the protracted dry spells and real droughts the combined water storage shrinks, probably by first one dam and then another, together with their respective irrigation areas, dropping out of the irrigation procedure.

When only enough water for stock remains in a dam, irrigation from that dam must cease. With the dry spell persisting, the owner will always be able to predict discontinuance of irrigation from a dam well in advance and thus be able to make his decisions accordingly.

There are yet other factors which may in dry times assist the farmer in the management of his waters for irrigation. For instance, and as with many Keyline layouts now in being, dams for the storage of irrigation water are often interconnected both above the dams and below them. The lock-pipe system, through the bottom of the wall of each dam, permits the rapid release of the water for either irrigating or for the transfer of water around the property. Thus the dwindling water of two or more dams may be reserved to the one irrigation paddock while withdrawing from irrigation the area normally serviced by one of them.

The same interconnected or chain-type layouts often involve dams, or even chains of dams, at two and sometimes three different levels, thus permitting the transfer to a lower dam of lessening water reserves to maintain full irrigation on particular paddocks. It has been found in practice that the more severe the drought the more marked is the tendency for the dams located in the higher parts of the property to run out of irrigation water and that the farther down the property one moves the more is water still available. One of the reasons is that the most favorable sites for dam construction are often associated more with the middle and lower slopes. Further, the filling characteristics of the dams generally improve toward the lower country. Of course if waste of water were to be absolutely prevented it would be necessary to know exactly when it would rain and how much, so that then irrigation could be arranged as would cause all dams to become empty at the one time and just before the filling rains arrived.

There are indeed a few places on the earth's surface where both the rainfall incidence and the amount to be expected are so consistent that water usage from farm dams for irrigation could be precisely planned. But while such weather forecasting is an unattainable Utopian dream to most landsmen they can still do a lot better with water from rainfall by using the logical methods of planning and design. However it does not involve avoiding the use of irrigation water because there might be a drought and he might run short of water. It is much better generally to "cash" his stored water at the first and every available opportunity. But use of available water is also more efficiently done it if is properly designed. The design of water use from various farm dams must reconcile several possible rainfall patterns. First, the future rainfall the next year or two may be such that the landsman cannot use up his stored water as fast as the supply is replenished. Second, the future rainfall may provide one or two runoffs and then a long, dry spell. Third, there may be no more runoff for a year or more.

In the first case, if irrigation water is not used on every occasion where it would be profitable, water has been wasted by not being used. In the second case runoff rainfall could be wasted if water flows off the property while some dams remain unfilled. Therefore the design of water use should provide for a quicker and larger draw from the last dam of any chain of dams. Then final runoff water would be trapped in that dam and not run to waste from its spillway if it were overfull. And if such a dam becomes short of water for its own irrigation area, water remaining in the higher dams can be used to service the lower irrigation area or to replenish the dam as required. In the third case of immediate drought the above design of water use will not have wasted water and the severity of the earlier part of the drought will be posponed.

These comments on dams and irrigation have been made on the subject of the assessment of farm water resources, so the progressive nature of the developments from Keyline water planning may be fully understood, and so that the safety of this manner of planning against the lack of information on the subject may be appreciated.

CHAPTER SEVEN

Practical farm water control

EXAMINATION OF WATER POTENTIALS. This discussion may proceed on the assumption that we are dealing with a property on which there is one or more complete water runoff systems, that is, a system in the sense of meaning an area of land on which there is a main ridge with its series of primary ridges and valleys falling from one side of it to, and including, the watercourse below. It is further assumed it is a property of 480ha of land which is of a medium undulating and smooth character that, while generally being short of suitably spaced rainfall, does have an appreciable runoff at completely unpredictable times, but usually once and sometimes three times a year. At the same time, the variations of any plan which may be necessary if the runoff were very much less, will be mentioned. Also it can be assumed that the only information, other than that from the farmer's own observations and experience, is that the mimimum annual runoff is only 51mm.

To begin with the minimum annual runoff represents 240,000m^3 (200 acre feet) of water. A minimum annual runoff of anything at all is a favorable position since the average annual runoff must be considerably higher, and the maximum runoff which may be expected once in say each three years, could then be very considerable.

The owners of agricultural land are more likely to go out in heavy rainfall than those who follow any other occupation and be more in a position to observe the high runoffs from the valleys of the farm and the state of the flow of creeks and other streams which follow on the rains. And that information, if the farmer has been thoroughly observant, is the most valuable likely to be available to him for planning the control of those waters.

Diversion channels for water control

The runoff from the property itself is, in the circumstances so far projected, quite significant on its own for development. The next point to consider is the flowin water of number 3 category. If the small creek flows intermittently and carries considerable flows after rainfall, it may bring to the property much more water than all the water of number 2 category, rainfall runoff from the farm itself. It is therefore examined for the possibility of either storing the water flow in large quantities at the nearest suitable site to the point of entry, which is the highest side, or controlling the water to divert the flood flow through the property and for storage at sites in some of the primary valleys selected for the purpose.

If the course of the creek through the property is somewhat rocky or of an uneven bottom with the stream flowing as a succession or rapids and small fall-overs, then the fall of all the land itself will most probably be sufficient to permit diversion of the water from the creek near where it enters, to fill properly sited dams. The fall given to a main water diversion channel for the purpose would be very much less than the average fall of the creek through the property and so, as the diverted water progresses down the property it has the result of leaving increasing areas of land below the channel and between it and the creek.

That feature is of great importance as it may allow of dams being constructed to store the water, dams which when equipped with a suitable sized lock-pipe system, will command areas large enough for low-costing gravity irrigation. Where those conditions obtain, the site for the control of the flood-flow of the creek and of the diversion channel itself becomes the first fixed sites and water-lines — the channel — of the plan for development. At that time a line could be levelled in with a dumpy or Bunyip level, to show where the diversion channel would lie. To do that a site would first be chosen for the level at which the water would be diverted. Such a site is discussed here briefly while more details are given in other chapters.

As the diversion of the creek water would be accomplished by a dam and spillway at the most suitable spot, the level of the spillway would be the controlling level for the pegging of the diversion channel course down through the land. In undulating country such a channel should be given a set rate of fall, and 1 in 300 has been found as steep as is necessary in almost any circumstances.

Water storage sites located by diversion channels

Because the level point of diversion is to the side and on the creek bank and not on the bottom of the creek, the pegs set with the aid of the level instrument will be up on the land and progressively leave the creek farther as more pegs are set. Pegs should be placed at a fixed intervals apart such as 15m or 20 metres, and so will soon arrive to cross the lower end of the first of the primary valleys. The position of the first valley is the least likely one to be suitable for a storage site having land for irrigation below a dam constructed there. But it may happen to be a very good storage site from which water could be gravitated to land farther along. If that is so, then a decision should be made as to the most suitable wall height for a dam if one were to be constructed there later. A wall which will hold a depth of 6m of water above the level of the valley floor at the wall site is generally very suitable for medium undulating country. Where the land is much more gently sloped, a lesser depth of water may be decided on. Some of the critical matters to be considered in the selection of sites for storage dams and the heights of the earth walls are discussed elsewhere.

If a 6m dam is envisaged, then when the pegs of the diversion channel site-line reach a position in the centre bottom of the primary valley, that level point will represent the top water level of the suggested dam. A special distinguishing marker, such as a longer peg or steel fence post, could be set to mark the spot. At that stage a true contour line may be pegged around the primary valley to mark out the complete top water line of the dam. At the two places on either side of the valley where the pegs reach 6m above the valley floor which lies between them, would lie the wall of the dam. Pegging out immediately permits a full appreciation of the position as a dam site, as the landowner can now make a quick estimate of the area enclosed by the contour line round the valley and the line of the wall.

That area, multiplied by the 6m depth of the dam and also multiplied by a factor of from 0.3 to 0.4 is the approximate capacity of the dam in cubic metres. With a little more estimating the farmer can also know the volume of the wall and the approximate costs of the earthworks of the dam.

Continuing from the wall peg placed on the far side of the valley and resuming again the same rate of fall, the line is continued in that manner to, through and beyond the last primary valley of the system. On the way each valley can be appraised as a site for a storage, the areas below are observed to estimate the area which would be suitably positioned for gravity irrigation; critical distances, such as the increasing length of each successive primary valley lying below the pegged line to the creek below, are noted, and a great deal about his land is learned by the farmer in the course of pegging the "new" water line.

Every shape and form of the land is emphasised by just that one line, and often to the extent that the owner for the first time will realise something of the hidden potentials of his own land.

Choosing the first water storage site

To continue with this description of water category 3; assuming there is more than enough water flow to fill all the suitable dam sites which have been located, the owner is now in a position to choose the side to be used first. Any position on the pegged line can be selected for the purpose, and the main consideration may be with a site associated with an area which can be most quickly prepared for irrigation, that is the dam site and land associated with it, which will return its cost most quickly. As lines for water control and storage sites themselves are critical because all other factors of a plan are decided in relation to those basic lines, the next information required will relate to another aspect of water control, namely the more exact position of the irrigation area associated with the dam which is likely to be constructed first.

As the dam is to be 6m deep, a peg position in the valley below is determined 6m lower than that channel line peg which represents both the spillway level and the irrigation channel which would carry the water for irrigation. The irrigation channel falls with the land, the direction of the creek fall, as does the diversion channel itself.

From the irrigation channel peg, a line falling at 1 in 300 is pegged with the aid of the levelling instrument and continued right around the primary ridge at least to and just beyond the next primary valley. Since all the land below that line to the creek below will be commanded by the irrigation channel line just pegged, an estimate can now be made of the land as an irrigation area.

The main diversion channel first located will not only carry the flows diverted from the creek, it will also collect the rainfall runoff from the land of the farm which lies above the channel and throughout its length. The water from the farm is in this instance of such proportion as to warrant control and storage for use in irrigation, even it if is assumed that the soil improvement program and then later, the improved soil itself, will take a quota of it. Therefore the next information which is required is the relationship between the different Keylines of the primary valleys throughout the system. In this examination those primary valleys which have shapes of likely significance for water storage are the principal ones considered.

Inspection may start from either end of the system or it may be undertaken from a particular primary valley which appears to have the best shape for a dam site. The Keyline would then be levelled in and pegged around the valley to represent the water line of a storage as already described for the dams on the creek diversion channel. Estimates would be made on the spot of the water holding capacity, the amount of earth to go in the wall of the dam and the cost of all earthworks. Then a second line for a new diversion channel would be pegged from the Keypoint of the valley and rising at the set gradient towards the rise of the country, so that the full extent of the induced catchment area could be determined. Rising toward the rise of the land, the line therefore falls from the high country to the dam site, and becomes the line for a water collecting channel for the dam. When the pegged line reaches the next higher-in-the-series primary valley, that valley again is examined as to its suitability for water storage. If entirely insignificant for the purpose the pegged line may be continued around the primary valley and on to the other valleys which can be examined, as in the earlier example.

Because of the rising relationship of the Keylines of the primary valleys, the pegged line will probably cross around the valley which has no storage site at a point below the Keyline of the valley. The pegged line thus stays below that Keyline and does not restart and continue from it. The worthwhile result is that the line stays somewhat lower down and so has the maximum area above it to supply runoff water to it and finally to the dam.

If a later primary valley has a shape of significance for water storage the pegged line is terminated at that valley and at a point which is usually somewhat lower than the Keypoint of the valley. The Keyline of the valley is then pegged as a true contour to disclose the details of the possible storage site, and if it is satisfactory then a second rising line is started from the Keypoint of the primary valley and continues into the rise of the land as before.

All the primary valleys are examined for possible storage sites at their Keylines and for the relationship in height of their respective Keylines. A principal consideration is that the final results should be two or more interconnected storages in which the overflow from the highest dam is caught by the water-collecting channel and directed to the next lower dam, and so on throughout the chain of dams. If on occasions the sites as first selected do not quite conform to the pattern, then a little adjusting to raise or lower a dam site by 0.3m or so, and the gaining of a little height by slightly flattening a drain throughout its length, is resorted to. An unnecessary break in the chain which may prevent the overflow of a dam reaching the next dam farther down-land, or a break which divides what should be one chain into two smaller chains, cannot but reduce the efficiency of the system.

There are three likely avenues of gravity use in irrigation for the stored water of a chain of dams located on that pattern, as well as any method of spray irrigation where it is appropriate. First, if there is sufficient land lying below the lock-pipe outlet of the dam and above the creek diversion channel, which lies below the higher system, that land is irrigated via an irrigation channel, as earlier mentioned. Second, and a little less likely except in "short-slope" country, the water may be directed to and through the creek diversion channel bank on the lower side of the primary valley which contains the dam. The water may then be used to irrigate the strip of land lying immediately below the creek diversion channel and above the irrigation channel which employs the water of the dam on the creek diversion system. (See figure 10, page 82.) Third, the water may be used to provide extra water, where it may be required for an area belonging to a dam on the lower chain, by being released to maintain the water of such a dam and so increase its irrigation capacity.

The water lines for the two type of channels described now become the principal planning lines for the full development of all the land of this water system of the property.

Plans for every other feature are fitted into the water-lines design as discussed throughout this book.

Channel lines remain the same for less water

Another aspect of the planning relates to the lack of reliable information on the water available and the approach to the undertaking if there were considerable doubt about the water potential, or the water was much less than is indicated by the earlier comments.

The owner would first have to decide from his own observations and local knowledge whether the creek flow was such as to fill satisfactorily at least one dam instead of the whole series of them. Even if he were sure only of the water for one dam, the planning and pegging would still proceed exactly as before, except that, while every likely site is appraised as has been described, only one site would be selected for dam construction.

That site would be far enough down the property for suitable land to lie below it for irrigation if possible; the site would be chosen for economy and one capable of quick development and good financial return.

Similarly with the appraisal of the sites higher in the primary valleys, the owner may be confident that the water available could be managed by only one dam.

The appraisal of the various sites in that case would be practically the same as before. The particular factor to be noted is that the principal water lines which will influence the whole planning layout are the same for both sets of circumstances.

Later, when the lesser scheme is developed and working well, it may be found there is more water available than was expected and which the limited layout can use. Then it is a simple matter to add more storages on either water line channel, by fitting them into their appropriate relationship with the particular channel. The critical factor in the rearrangement is that the top water level of any new dam should coincide with that of the diversion channel on the down side of the valley, as that is the level at which the spillway of a new dam must join up with the diversion channel. The diversion channel as it approaches the

KEYLINE FLOOD-FLOW
FOR FLAT LAND IRRIGATION

Aerial view of a Keyline flood-flow project in New Caledonia showing the diversion channel, left, the large dam, centre, and the start of the flood-flow irrigation channel and bays, right. The dam is maintained full for most of the year by a small constantly flowing stream turned into the diversion channel. Irrigation rate is upwards of 8 ha an hour.

PLATE 24

Flat land irrigation. Shows the first test of an axial flow pump for flood-flow irrigation. Lifting the water only a metre or so the pump has a maximum capacity of 3600 m³ (800,000 gallons), which amounts to more than 3500 tonnes of water each hour.

PLATE 25

The same flow is used to test a new irrigation channel bank. The speed of this flow illustrates the work force gained by using a large flow of water. Irrigation rate is 6 ha an hour with one-man control.

146

PLATE 26

Steering bank for flood-flow. The first of two sidecasting runs done with a small Caterpillar tractor in the construction of a steering bank.

PLATE 27

Irrigation land on the same flood-flow project a year later. One grassed-over steering bank in the left half of the picture runs down the maximum fall of the land.

Diversion channel. The third sidecasting run in the construction of a water diversion channel which required 31 runs to complete.

PLATE 29

Portion of the same channel on completion. This diversion channel has a fall of 1 in 300 and carries a flow of 13,500 m^3 (3 million gallons) an hour. It transports the flood overflow of a creek dam to fill four more dams on Yobarnie.

KEYLINE FLOOD-FLOW IN TEXAS

PLATE 30

A new dam on a grazing property in the Texas Panhandle constructed to provide water for flood-flow irrigation. Water will be 7.3 m deep above the 610 mm lock-pipe system. The valley above the wall is very flat and the area of the dam is large and extends 0.9 km from the wall. Rate of flow for irrigation will be well in excess of one million gallons per hour. The dark area in the distance is the start of the irrigation land which extends to the right, out of the picture. The steering banks had not been constructed when the photo was taken.

Portion of a very large diversion channel 6.4 km long which collects rainfall runoff for the dam in the picture on the facing page. In dry times water will be pumped into the channel from a creek dam of large capacity, to keep the higher placed irrigation dam filled. The creek dam also has a 610 mm lock-pipe system installed. Two more dams are planned, one located on the present diversion channel and another beyond the dam opposite to which the diversion channel will be extended. Irrigation rate will be 10 ha an hour from each dam. Rainfall in the area amounts to only about 406 mm a year average.

Shows the experimental use of placing fences on steering banks for hillside irrigation on Yobarnie. The purpose was to subdivide an area of irrigation land into many small paddocks in such a way that during irrigation, paddocks carrying stock could be left unirrigated for the time being. The irrigation channel is 6.4 m up the slope from the post in the foreground. This steering bank fence joins up with another fence along a second irrigation channel lying just above the cattle. Note earth stops placed in the drain formed during construction of steering bank.

FENCING ON THE STEERING BANKS

PLATE 33

A large steering bank now grassed over in a flood-flow irrigation project in New Caledonia. The fence on the steering bank is for the control of cattle.

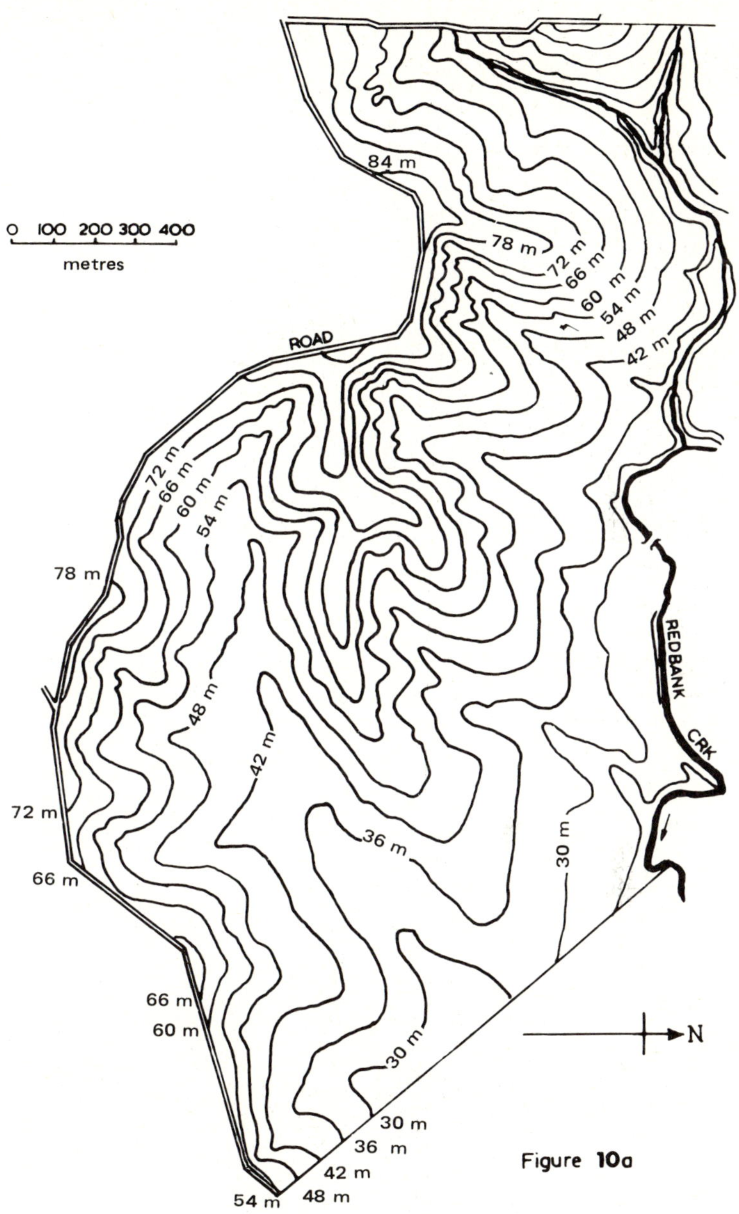

A large part of the contour map of Yobarnie. The road on the southern boundary is located along the main ridge. Note the primary valleys and primary ridges falling to Redbank Creek to the north. Contours are at 6 m intervals.

153

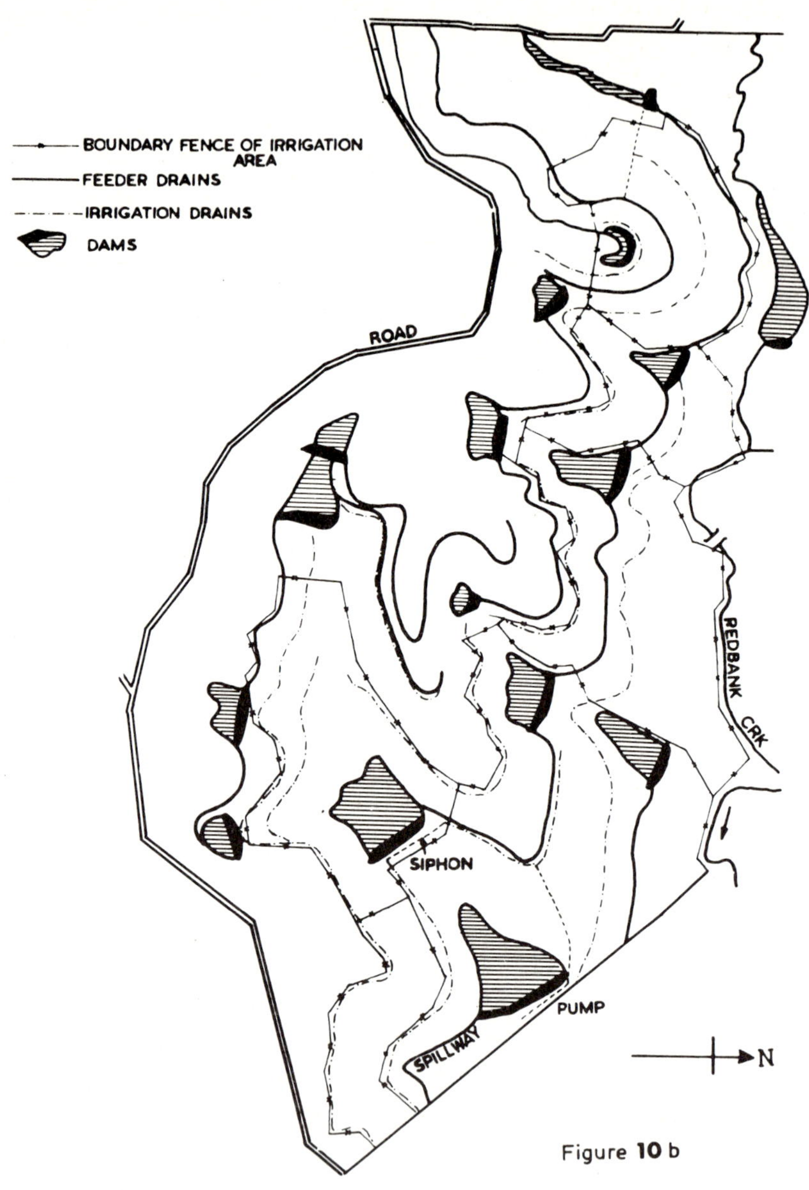

A functional plan of the same area shows the diversion channels as unbroken lines and the irrigation channels as dash-dot lines. Note that the chains of dam are also interconnected below by the irrigation channels. Compare this with the aerial photograph of Yobarnie (see Plate 5).

new site from the creek is thus above the top water level and needs only to be opened to allow the water to flow in.

Relationship of creek and channel falls

The feature which permits the use of the flows from a creek is a suitable relationship between the fall of the creek and that of a diversion channel. On a map of the land, the contours after crossing over the creek would first hug the creek course and then swing outward and farther from the creek as they progressed.

In other circumstances the fall of the creek may be too flat and the contours of the land remain too close in along the creek for the scheme to work. A site for a large storage is again sought near the creek's entrance to the farm, and from which the water may be pumped to a channel from which one or more higher storages may be filled, and in a similar manner as with the gravitational set-up. Then a pump of suitable capacity may be set up permanently to lift the water from 3 to 15m (10 to 50ft) higher according to the slope of the land. For instance if the average slope of the land, main ridge to creek, was 1 in 20 then a pump lift of 12m (40ft) would permit the gravity irrigating of a strip of land more than 120m (400ft) wide after allowing a 6m (20ft) height loss for the pump-filled dam to release the water through its lock-pipe system. If, on the other hand, the storage requirements were satisfied in a dam 3m (10ft) deep a strip of irrigation land 180m (600ft) wide would be provided.

In yet other circumstances the size of the creek storage may be much larger and be of maximum capacity as compared with creek flows; then the full irrigation potential of the situation may be satisfied by pumping into a channel for directly irrigating; and very large volume pumping is now quite practical and economical in such circumstances.

The lines of diversion channels are for property planning

Whenever it becomes necessary to decide how best to control and use farm water, and whenever channels for water diversion and for irrigation are involved, the lines of those channels are likewise the lines which most influence the planned development of all the land in their water systems.

Thus far only the water catchment relationships in one half of a land-water section has been considered; that is the one side of a main ridge and includes its primary valleys and primary ridges, down to and including the watercourse below it. When the examination is widened to the complete land-water section which embraces all the land lying between two creeks and including the creeks themselves down to their junction, new water catchments and relationships intrude, with wide opportunities for profitable water deployments.

It may be that two creeks enter the land and both carry water after rains. The height relationships of them at their entry points to the property may indicate various possible water transfers. The general fall of the creeks and of the land may allow of the gravity transfer of some of the water from one creek catchment to a storage on or near the other, to increase a supply to more economically usuable quantities; or all the land most suitable for water applications may be associated with the creek or the catchment with the smaller supply. More examination than was entailed in the first example will probably be necessary; Surely on that and other occasions of still wider scope, would not a good contour map be of outstanding value?

With the aid of a good contour map and the knowledge a farmer has of his land, the various contours of the map could first be considered as if every contour line of the map was instead a line which had the rate of fall and the direction of fall which might be required for water transfer. Then by following selected contours from the creek through to the other one and in reverse, all the possibilities of the situation would gradually unfold themselves; and what surprises and absorbing interest the occupation generally creates! Completely unsuspected relationship of water and land soon appear and often one item only may, when developed economically, add more to the value of the area than the property's original price.

Of course the contour lines on a map are not the falling lines of channels for water transporting, but, because the rates of falls for all types of channels, including those of the natural creeks and stream courses, logically tend to become less and less as the country being considered is flatter and the vertical tolerances dwindle, the examination of the contour map for such discoveries is a very practical approach. For instance, the general fall of channels in undulating country may be 1 in 300 but if that rate of fall were to write-off an undertaking because too much vertical height were lost, it would be an obvious step to lessen the fall to 1 in 400 or to 1 in 600 to gain the critical height. But on flatter land there may be no such rate of full fall as 1 in 600 because the fall of the country itself may be only 1 in 1000 or 1 in 3000, so an appropriate rate of fall for a channel has to be less than that of the country; and all those various considerations are soon obvious from a good contour map.

When it comes to the disclosure of suitable water storage sites on the larger properties of more gently undulating land, a contour map may reveal outstandingly valuable locations which an owner has not noticed in decades of property management. If the chief asset of land is water surely the property owner must endeavour to control its flow for his property.

NEW LOOK AT FARM WATER RESOURCES

While it is not possible to describe in the course of one volume all of the even more commonly met modes of water occurrences and availabilities and the manners of using the water within a land-water section as designated, there does emerge from this brief recital a few fundamental, practical and common sense principles and guides which will apply for widely differing circumstances. But to start with, the farmer's own land should now be looked at anew and older notions and conservative and conservationist egotisms and illusions about water discarded.

Throughout the whole of this continent water flowing off the farms and the grazing lands should be looked at first as money not staying where it rightly belongs. Then the landowner, examining his waters afresh, should realise that his most valuable potential water resource will provide by its own development and use the first basic details and lines for the planned development of his whole property and for the re-creation of its landscape.

It is certainly obvious enough that water flowing to the farm from a catchment larger than the farm itself should be considered in the first appraisal of farm water resources. Particularly so since it may not only supply the first permanent development lines but it may dominate the whole improvement and working program. The first place to seek to control this flowin water must surely be at or near the point of its entry to the property. And if it is not practical to do so there, control points are then sought elsewhere by following the stream course downwards. The higher up on the landscape that water can be made available the more valuable is the water, and the greater is the area of land which lies below it and on which it may be used. For farm runoff itself, category 2, the Keylines of the primary valleys mark the water line of the highest possible sites in the valleys for a dam to store such water, so they should receive consideration on the basis already described. While on occasions one dam site may be so outstanding in its advantages as to dominate the whole course of development, it would be only rarely that the great advantages in efficiency of the interconnected system of dams should be broken to take advantage of a more favorable site for one of them. Moreover there is usually a measure of adjustment possible which can be exercised in favor of both the dam site and for keeping it in the chain of dams.

Then there are occasions where a landholding has only one primary valley suitable for building dams, the other valleys, if any, having no shapes of any such significance. Two or more dams may then be constructed in the one valley, stepped one above the other. In that instance diversion channels would be used to bring in the water to each dam from the area which could best supply it or even from both directions if necessary. The overflow from the topmost dam may be directed to the next lower one and so on throughout the series, the valley, with its own dams and their channels, then supplies the planning lines for the whole property because of its predominating water influence.

But putting in dams in stepped fashion in a particular valley and employing an interconnected series of dams in another set of circumstances, does not make the two schemes competitive. It is simply that water overflowing from a higher to the next lower dam of a series of stepped dams, loses height of from 4.5 to 15m at each overflow. It is acceptable in design only where it is unavoidable, and especially so when compared with the loss of height in the overflow between dams in the chain of only from 381 to 1270mm. But there is a place somewhere for every useful scheme and idea to serve its purpose in the the development of the water resources of the farm and grazing lands.

This chapter has stressed the need for the proper assessment of the water resources of each individual farm and grazing landscape but has more or less shown, that, with the present lack of knowledge of all the relevant factors, it is nearly impossible of accomplishment. What is the answer? It is suggested that the only practical and worthwhile answer for the farmer and grazier to the problem of water resources assessment lies in the progressive and perfectly safe nature of the methods of planning outlined in this and other chapters of this book.

CHAPTER EIGHT

Keyline flood-flow irrigation

IN THE KEYLINE FLOOD-FLOW system for flat lands, water for irrigation is released, diverted or pumped from the source of supply into the start of an irrigation channel which lies across the fall of the land. The channel may be designed to have a fall from the source of supply or to be "dead flat" on a true contour and in accordance with the circumstances on each individual property. The guided and controlled large stream of water is next released from the irrigation channel through watergates into irrigation bays. The irrigation bays are formed by earth banks extending from the irrigation channel down the maximum, albeit gentle, slope of the land. With all the other watergates closed, when the irrigation of the first bay is completed the watergates from the irrigation channel to the second irrigation bay are opened and those to the first bay are closed.

Irrigation proceeds in that manner by releasing the full flow of the water in the irrigation channel into each individual bay in turn. Finally the watergates to all bays are again closed except those to the last bay which remain open after the supply is closed off so as to use and to drain the water from the full length of the irrigation channel. The following day all watergates are again opened to ensure good drainage. Should heavy rain occur between irrigation cycles, the open watergates will ensure an even distribution of the water from the rainfall runoff.

From that description the Keyline flood-flow irrigation system could be likened to the more orthodox border check method which is used widely in government-controlled irrigation districts. However the essential differences between Keyline and government-controlled systems are very great. First, the available flow of water for flood-flow is not limited by the need for measuring or by other restrictions on the flow rate, as is the water for irrigating in the public scheme. Thus the flood-flow water flow rate may be up to 30 times greater than in the public scheme and as a consequence of such nearly unlimited flow rate, the irrigation bays or waterruns may be 20 to 40 times larger. With such large bays, land preparation is totally different and instead of relying on extensive and expensive levelling and grading work to ensure an even spreading of a small stream of water, as with border check irrigation, the large flood-flow streams plus Keyline pattern cultivation, accomplish it at around 1/10th of the cost.

But perhaps the greatest difference of all between the two systems lies in the the philosophical attitudes to the irrigated soil. Whereas the more orthodox practices treat soil as if it were a static substance but one which could deteriorate, Keyline regards any existing soil as essentially the home of an organic community of living things which can be improved to a much higher degree of fertility. The opposing principle leads for instance to the practice of even reducing the small flows of the border check system to hold the water in contact with the soil for longer periods, so a predetermined amount of water is forced into the soil. Thus soil may be covered with slowly moving water for many hours during each irrigation, to the great detriment of soil fertility because of saturation and root suffocation.

On the other hand on a Keyline irrigation project the basic aim remains that of developing the highest fertility in the soil. And a really fertile soil, when it is somewhat dry, will literally gulp water almost as fast as it can possibly be applied.

So when soils for flood-flow irrigation are initially infertile, as they almost always are, and as a consequence do not absorb water rapidly, then that type of soil is chisel cultivated before irrigation in such a way that the soil will then quickly take in its quota of about 51 to 64mm of water from each irrigation. That type of cultivation is done at appropriate times and according to the requirement of the individual soil. By providing, as it is designed to do, improved soil aeration and better living conditions for the soil life, the particular soil immediately starts to change and with continued care, soon reaches a condition of high fertility. Then with the new sponge-like structure of fertile soil it is capable of absorbing water rapidly, and Keyline flood-flow irrigation continues to control the water and irrigate the land much faster and at a considerably lower cost than any other fully controlled irrigation procedure.

The matter of the water intake rate for irrigated soils, particularly in government irrigation projects, is the subject of many tests and experiments and frequent reference in journals on irrigation.

We may read that a certain soil when dry "takes in" 13mm of water in half an hour but very little more in two hours; perhaps that in 10 hours it still won't absorb 78mm of water.

In Keyline we are not concerned with such times but in how much water the soil will accept in the first 15 seconds, half minute, or 10 minutes, and therefore with what techniques of soil treatment are necessary to make a dense soil absorb water quickly. Any dense soil can first be mechanically treated to cause it to take in 51mm of water quickly. The treatment, when continued for a little time, will soon promote a new fertility in the dense soil and alter its structure to the sponge-like aspect of highly fertile soil so that later and with little more treatment, it will take in water almost as fast as it can be applied. A sponge plunged into water is hardly any wetter after 10 hours' immersion than it is from 30 seconds' or from 10 minutes' immersion, and so it is that the sponge-like structure of a highly fertile soil permits the very rapid water application rates of the Keyline flood-flow irrigation.

SOURCE AND DEVELOPMENT OF WATER SUPPLY

Some sources of very valuable water supply on farm and grazing lands may be obvious while others may not be so, and can even be completely concealed and unsuspected. The sources of supply for flood-flow irrigation are similar, generally, to those which may be available for other systems or irrigation except that because of the necessary flat land associations, storage dams are generally much larger and tend also to be shallower.

They are first dam storages which may be filled by runoff rainfall. A storage dam, or dams of suitable capacity may be constructed and filled with rainfall runoff water from the dam's own natural catchment, or runoff from a sufficient outside catchment area may be diverted to fill the storage. According to the local circumstances of rainfall runoff, land shape and the related capacity of the storage, and to the area of land which may be available below or near the supply for irrigation, a lock-pipe system of suitable size is envisaged which will eventually release the water from beneath the earth wall of the dam as required for irrigation.

The widely varying circumstances on individual properties will produce an almost infinite variation in the shape, type and capacity of such storages and thus affect their rational uses. But those farm storages will probably provide the water for the greatest irrigation potential of all and could lead eventually to the development of more irrigated land than from all other public and private water sources combined.

Another type of outlet for irrigation water additional to a pipe beneath the wall may be used if such a dam is constructed to tap a perennial stream or which for some other reason can be kept constantly filled. Thus a watergate may be installed in a position to control and release for flow the topmost foot or so of the water of the dam. Hence such a storage would have two types of outlet: (1) A lock-pipe system through the bottom of the wall, which could supply water to an irrigation channel positioned at the lower level, and (2) one or more watergates to supply water from the top of the dam to an irrigation channel on the higher level. Again in the circumstances when the stream flow is less than usual, the land serviced by the lower irrigation channel can be maintained in a fully irrigated condition, while the land serviced by the higher irrigation channel receives irrigation only when the dam is filled to a sufficient capacity to supply water through the higher placed watergates. It is a simple matter to release water from near the top level of a dam via watergates at low cost and in a practical manner. In other conditions a high capacity, low head pump can be used.

Second, a perennial stream of sufficient flow capacity may supply, by diversion or pumping, a minimum flow of, say, 2250m^3 (hold a million gallons) an hour. Such a direct source of flow is sufficient to supply a maximum iririgation area of up to 320ha (800 acres) on continuous irrigation for eight hours a day on a minimum irrigation cycle of 10 days, and assuming that 570m^3/ha (50,000 gallons) of water was used at each irrigation.

That source of supply can also be associated with a constructed storage and it is to be noted that very large farm and grazing land water storages are most practical on some of the flatter areas. Such a storage dam can then receive continuously the full flow of this stream day and night up to the amount that can be pumped or diverted over a full irrigation cycle. Thus, if the dry season irrigation cycle is 10 days, the hourly flow multiplied by 10 days of 24 hours would provide to a large storage available water for each irrigation of 550,000m^3 (120 million gallons), or sufficient for fully irrigating about 960ha (2400 acres). A dam of that capacity on, for instance, a medium-sized grazing property, can be well within the financial resources of the owner.

Now a potential of such size can be discovered and determined quite quickly. It can then be decided to use a flow rate of 11,000m^3 (two million gallons) an hour and thus to irrigate at the rate of 16ha (40 acres) an hour, whereupon 60 hours of irrigation would complete the 960ha. The size of the lock-pipe system to release the irrigation water from below the wall of the dam and/or the capacity of the watergates to release the topmost foot or so of water is then decided by the requirements of the designed flow rate.

It can be seen that the same conditions also apply if the flow available is only 90m^3 (25,000 gallons) an hour; but then, on the same basis, the maximum irrigation area which could be supplied is 48ha (120 acres). In such circumstances the capacity of lock-pipes or watergates of a storage to control the flow can be designed to release the water which was accumulated from 240 hours of flow and complete the irrigation in one day. Such a storage has the very modest capacity of 27,000m^3 (six million gallons). Why in such circumstances should an irrigation system be installed at very much higher cost and be designed to use the water at its available rate of 90m^3 (25,000 gallons) an hour and thereby complete only 20ha (50 acres) in 100 hours of irrigating?

And there is this to be said against the present widespread practice of designing small farm spray irrigation layouts on the basis that the time occupied in spray irrigating should extend over the full irrigation cycle, and so it is believed, thereby reduce the capital costs of the installation. But a small or one-man farm irrigator should not be occupied for 10 days in irrigating a small area when the minimum and critical irrigation cycle is 10 days. A small area irrigator certainly should not be a man who is continually occupied in applying water to land. Instead he should be a manager of land on which a little irrigation is practised, and he needs usually not less than five of the 10 days of the irrigation cycle to manage other aspects of the land, namely, the soil, grass, stock fences and crops.

To get the best management and the most rewarding results from small area irrigation, whether it be spray irrigation or one of the many systems of flood and flow irrigation, the complete irrigation should be finished in no more than half the time of the minimum irrigation cycle. If at all practicable the layout and design should provide that the irrigation can be finished in one day.

If the critical irrigation cycle and the time occupied in irrigating are 10 days each, what happens when it rains and completes the proper wetting of the whole area in a few hours or a day or so? Does irrigation start again the day following the rain, or the next or some days later? If irrigation does not start the day immediately after rain then the paddock or portion of the area to be completed on the 10th and last day will be without water for some days at a critical time. And is not the most beneficial result of irrigation obtained by the application of water in critically dry times but before it becomes critical for the irrigated land?

The third source of water supply is underground which, by bores, is brought to the surface and accumulated in a storage. Such storages will generally be of the flat land type. For example, a large valley storage, a contour dam which is very suitable on slopes of 1 in 50 to 1 in 100, or a large ring dam which is suitable on the still flatter lands.

Again a lock-pipe system of calculated capacity may release the water into an irrigation channel. On some occasions watergates are again suitable for releasing the top metre or so of the surface water of the filled storage, and thereby a few feet of additional height are gained which, on very flat land, may bring into the scope of possible irrigation an extra strip of land 305m or more wide for each vertical 0.3m gained, or 1m in 1000m.

CHAPTER NINE

Channels — some observations

WATER HAS BEEN TRANSPORTED from place to place by channels in the earth since time immemorial, and it is still the cheapest method. But as we have said, there are two prices to be paid for water, the second being the cost in water itself.

The second cost in water lost in the transport of water via channels becomes continually higher as the distance which the water is transported becomes greater. The enormous loss of water from government irrigation supply channels is a matter of continual concern. But the solution to the problem of water lost by seepage and to a lesser extent by evaporation is enormously costly. It could involve the conrete lining of hundreds of miles of main supply channels or, alternatively, the replacing of channels with huge pipelines.

But there are no such problems with the use of channels for transporting farm waters around the farm. Here the channels are cheap to construct and they don't waste water, there is practically no cost in wasted water because the distance the water travels is, by comparison, insignificant. Moreover its spread of travel is considerably faster than the slow movement of water in government supply channels so that the ratio of seepage and evaporation loss, without regard to distance, is also probably a great deal less. There are as well, other farm-favoring comparisons. For instance, when the largest farm channel, the diversion channel, is carrying its greatest flow of water it is sure to be raining heavily. Where then are the seepage and evaporation losses? On the other hand the government supply channels are very large and carry their greatest flow in hot, dry weather when all farmers would like to irrigate their land.

When a farm irrigation channel conducts farm-stored water for irrigating, the water has been released direcely into the channel at the quick turn of a tap, or pumped, and is spilled out from the channel to irrigate the land after travelling only 18m or maybe less than 6m, while at the end of irrigating the distance of water transport would scarcely have reached 732m. But even that same small channel, when used for reticulating the water of the government irrigation scheme from farm to farm, is very long and the water travels slowly. In addition, seepage is a universal problem of the large irrigation area itself and has caused many problems of swamp creation, salting and waterlogging even to the serious stage where whole productive areas may be made entirely unproductive. When the farmer is using his own developed water resources no such problem can exist. If there is a little seepage water, it will in all probability be only slightly less valuable than the controlled irrigation itself.

In that and in every other phase of the use of water for agricultural use, no other set of general practices can approach for sheer efficiency in the use of all the available water the properly planned development and management of the farm's own water resources.

It is not the purpose here to present all aspects of channel uses on farm and grazing land, or to provide a working manual from which could be determined all the aspects of volumes of flow and related channel sizes, shapes and rates of fall. For the present, and except where we give details of specific cases, such matters can be left until later. Moreover, there are many circumstances where those details can best be determined on the development property itself and if necessary with the aid of engineering advice with which such matters can be precisely determined.

It is however the intention to illustrate here the forms and the practical uses of channels so that their general types and particular uses can be appreciated by the landowner in the possible application of the types suitable for his own land. There should also be sufficient information herein to assist the owner in the intelligent consideration of any advice which he may seek on these matters.

In the development of the water resources of farm and grazing land there are two principal uses for channels: first, for the diversion of direct rainfall runoff, stream flow and pumped water into a dam for storage and second, for carrying water to a specific area for irrigating the land.

The various forms in both types of channel are dictated by the shape of the land through which they pass and by the volume of water flow which they are to carry.

A diversion channel used for collecting rainfall runoff from the steeper country to fill the highest located dams, those placed at the Keylines of the primary valleys, is a very different undertaking to that of a diversion channel in flatter areas.

In the first case the channel on undulating land would consist of an earth bank and the channel against it from which the earth was taken. It would be constructed generally by a bulldozer equipped with an angle-and-tilt blade or by a power grader, sidecasting the earth from the land immediately above to form the bank. It may be completed in from 4-10 sidecasting runs. The rate of fall of the channel would be 1 in 300 and it may carry a flow of $4500m^3$ (one million gallons) of water an hour. It would carry only a small part of its capacity in the excavated portion, the main part flowing against the earth bank. In total width from the lower edge of the bank to the upper edge of the cut, it may be only 4.5m. In appearance and cross section it would as shown in *Plates 18 and 19,* and the cross section in figure 11 below.

Cross section of a diversion channel in undulating country. Fall 1 in 300.

In the second case, a channel to carry a similar amount of water but in much flatter country could be described in the same way, but a much greater number of sidecasting runs would be necessary for its construction. Also its size and appearance would be different.

According to circumstances its rate of fall may be anything from 1 in 500 to 1 in 5000. It would carry more of its water capacity in the excavation from which the bank was dug but well over half of it would be held by the bank. It may be built by sidecasting as before but requires 32 or more runs to complete. More conventionally it would be constructed with the straight blade of a bulldozer working back and forth at right angles to the channel and from above it. The width would be from 12-18m. The appearance and cross section are illustrated in a photograph, *Plate 28,* and figure 12 above.

Figure 12

Cross section of a diversion channel in flatter country. Fall 1 in 500 to 1 in 5000.

A second comparative illustration is of two irrigation channels, first the excavated channel of near present maximum capacity for Keyline pattern hillside irrigation and, second, a surface channel of about half present maximum capacity for Keyline flood-flow irrigation on flat land.

The picture (*Plate 20*) and the cross section in figure 13 opposite, shows the first irrigation channel. Its width is 1.2m (48ins) and depth 0.6m (22ins). It falls at the rate of 1 in 300 and the rate of water flow is $1600m^3$ (360,000 gallons) an hour. The water is delivered directly into the channel from the 305mm (12in) lockpipe outlet of a storage dam.

Figure 13

Cross section of channel for hillside irrigation (Keyline Pattern). Fall 1 in 300.

The channel is excavated with a tractor-drawn delver I designed, and a small farm ditcher. The ditcher is used after each run with the delver to move the excavated earth away from the channel edge and to spread it. In that type of channel it is more or less critical that the lower edge of the channel coincides with the accurately pegged falling line, therefore the excavation is made just above the line.

For irrigating, the water is backed up with an irrigation flag, which forms a channel dam or stop, and spills down over the sloping land below. When the area is watered the flag is lifted causing the water to flow on and to be again backed up and spilled over by a second flag while the first flag is replaced in the channel ahead of the second flag. Irrigation proceeds in this manner away from the dam to the distance end of the irrigation paddock. Uniform spreading and soil penetration of the water is achieved by Keyline cultivation and soil management and by the type of control which the landholder can exercise over the flow of the water with his flag movements.

The second irrigation channel for flat country shown in the picture in *Plate 21* and the cross section of figure 14 is for Keyline flood-flow irrigation. The channel is 17m wide and constructed on a true contour with no fall. It is flowing a 4500m^3 (1 million gallons) of water an hour and has a capacity of over 7857m^3 (1¾ million gallons) an hour designed ahead for more water which is becoming available.

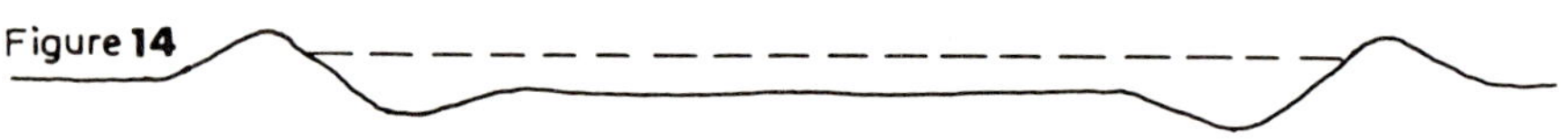

Figure 14

Cross section of irrigation channel in flat country (flood-flow irrigation) — No fall.

The pegged line for the channel is along the lower bank and the second bank is parallel above it. Each bank was laid in by sidecasting in two runs. The water is held by the banks and so it flows on the top of the land. Below the channel is a series of small earth banks which form irrigation bays of an average of about 4ha each. The irrigation water of 4500m^3 an hour flow is released on to the land of each bay in turn.

In the first of the two types of irrigation channel, the excavated channel for undulating country is limited in its capacity and therefore its rate of irrigating land, by two factors. In the first place it is designed for one-man use, and as it is frequently necessary for the owner to step over the channel and on occasion even to straddle it, the width thus has definite limits. In the second place it is designed for low costing construction by farm equipment with only the general addition of the delver, therefore the drain size is again limited by that consideration.

Although it has not yet been done there is no reason why the system could not be made larger and designed with bigger channels to carry twice the water for two-man control. Then the flags would be much larger and placed in position and released from the channel by a man at each end of the flag bar on each side of the channel. Such departure would be warranted in the very favorable cicrumstances where there existed an association of large-area primary ridges and a bigger capacity storage site in a nearby primary valley or where, from a large storage, a pump could lift the water economically to an irrigation channel set, for instance, not too much higher than the top water level of the storage. The object of both these setups could be the irrigation of 32 to 40ha a day.

The channel and flag method of controlling irrigation water, while designed specifically for Keyline Pattern irrigation for pasture and grain, also lends itself ideally to cash crop and orchard furrow irrigation with only the obvious slight variations.

With the irrigation channel for flatter land flood-flow irrigation the only apparent limits to the rate of irrigation are not those imposed by human considerations. Because the watergates by which the water is controlled and released in irrigating are only increased in number to each bay and not in their size, the farmer would not be taxed in controlling a $9000m^3$ (two million gallon) flow instead of only half that. The limitations are the obvious ones of water supply sources and the availability of a sufficiently large area of uniform and gently falling land.

As has already been said, earth channels are the lowest costing means of transporting water. In the development of the water resources of farm and grazing land and in the direct irrigation from those water resources, the appropriate channels must play a major role. It is of great importance also that the sites and positions for the various types of channel should be located according to the dictates of good design since channels may play the dominating role in the planning for the maximum development of agricultural land.

CHAPTER TEN

The flood-flow irrigation channel

THE TERM FLATTER LANDS signifies all land having fairly uniform slopes with a fall of from 1 in 40 to 1 in 10,000 or even flatter. It is to be noted that the cut-off point in the classification of undulating land and flat land can not always be precisely determined and direct experience on a particular property having borderline slopes may be necessary to determine the choice between methods of irrigating the land.

While flood-flow irrigation itself is suitable for such wide slope ranges, it becomes necessary to employ two distinct types of irrigation channel for the transport of the irrigation water to cover the full range of slopes.

On the steeper of those flat land slopes of from 1 in 40 to say 1 in 100, the irrigation channel is formed by a single earth bank about 0.6m high, which is constructed across the general fall of the land. The irrigation water then flows on the land on the upper side of the bank. The width of the irrigation stream will thus vary according to the volume of flow and to the local slope variations of the land immediately above the irrigation channel bank and throughout its length. Thus an irrigation stream of constant flow volume could vary in some circumstances from 6m to nearly 30m wide.

On lands where the slopes are flatter, the irrigation channel is formed by two such earth banks constructed in parallel from across the fall of the land. The irrigation water then flows on the surface of the land between the parallel banks. The distance apart of the banks which form that irrigation channel will depend on the volume of flow of the irrigation stream. The banks themselves may remain the same size for irrigation stream flows varying from 2-9000m^3 (one half to two or more million gallons) an hour. The channel varies only in the distance apart of the two banks.

It is to be especially noted with both types of irrigation channels for the flood-flow system that (1) the water flows on the surface of the land and not partially within the ground in a relatively narrow and excavated channel, (2) the width of the elevated channel is much wider than the irrigation channels for other flow systems, and (3) the irrigation water of the flood-flow channel, being on and above the level of the land, is available for low cost and rapid release into the irrigation bays.

Marking out the irrigation channel

The starting point of the irrigation channel is at the delivery point of the irrigation water. The delivery point may be (1) at the lock-pipe valve which releases water through the base of the wall of a storage dam, pond or lake, or (2) at the watergates which release the top one to two feet of water from a storage dam, or (3) at the delivery or discharge outlet of a pump, or (4) at the point where a diverted stream arrives on the area to be irrigated.

The general level point or datum for pegging the levels for the line of the irrigation channel then relates to the invert or bottom of the lock-pipe outlet, or to the pump discharge pipe, or to the horizontal channel frame or sill of the watergates which have been placed to release the top water of a storage.

The starting level of an irrigation channel which is to receive the water from a lock-pipe below the wall of a dam or from the outlet pipe of a pump can be taken directly from the invert of the pipe outlet, or from a few millimetres to 0.3m below that level. In both cases the water which flows into the channel is forced out by positive head or pressure. The flow into the irrigation channel from a lock-pipe beneath the wall of a dam has as head the depth of water above the outlet in the dam. Thus a depth of water of 6m (20ft) above the outlet produces a water pressure of 60kPa (8lb a square inch) and a depth of 2.4m (8ft) above the outlet, a pressure of 20kPa (3.2lb a square inch). The flow from a pump will also have a positive pressure.

In both circumstances, and for the same reason, it is possible to have the level of the irrigation channel at a height such that when the water flows it can build up and partly or even fully submerge the pipe outlet. The loss of flow will be small. But that does not apply to the release of water from the top of a filled dam through watergates. If the position of the irrigation channel is such that water soon builds up against the flow through the watergate then the flow is impeded and greatly reduced.

A recently designed overseas Keyline development includes a sizeable flood-flow irrigation project. The irrigation channel was sited on a level 381mm below the level of the sills of the watergates. The watergates were designed to release the topmost 381mm of the waters of a dam, which was itself kept fully filled by the water from a constantly flowing stream diverted for the purpose. The designed flow of 4500m^3 (one million gallons an hour), for the first stage of the project, was maintained through two 0.6m wide watergates. The diverted stream which was itself fully controlled by a watergate at the stream source some distance away continued to flow into the dam during the irrigation. With its large surface area the dam lowered only a few millimetres during the irrigation of 81ha of pasture. The top water level at the start of each irrigation, being 381mm above the sills of the watergates, was thus 762mm above the ground level of the irrigation channel to which the water flowed.

The level point at the start of the irrigation channel can be lower than indicated and that could only improve the water flow position. However it is usually desirable to have the irrigation stream as high on the land as is practical for flow.

On the other hand the pressure water from lock-pipe or pump flow may on occasion be raised 0.3m. That would involve higher banks on the irrigation channel for a portion of the length at the end near the water inlet, and the creation of a pond area to raise the water level the extra 0.3m to cause it to flow along the channel. Such a design feature does however, reduce the amount of water finally available from a dam as it becomes empty, by the quantity of water in the dam representing the extra height. Again in the circumstances where there is more water than is required to irrigate the land which is available below, then the extra 0.3m of height is very valuable in that it substantially increases the area of irrigable land. For instance, where the land has a general slope of 1 in 2000, the extra 0.3m adds a strip of land for irrigation 610m wide and for the full length of the irrigation channel.

However, with the starting level point decided, a peg is driven in to mark it and another reference level at the same height is marked and permanently pegged some distance away where it will not be disturbed or lost.

Pegs

It is a considerable advantage to use good pegs, which are of a height suitable for the surroundings and which have been painted for easy location and siting. It has been found that tomato stakes of 25mm square timber 1.8m long make very suitable pegs. The stakes are cut in two, or in three pieces if the grass is short, and the top 75mm dipped in a white paint.

Levels

One of two different type of levelling instruments will generally be used to level-in the irrigation channel line. They are the engineer's ordinary dumpy level and the transparent-hose water level. My own designed Bunyip level of the second type consists of a length of 60-feet of half-inch transparent hose with five feet of each end fitted into two specially shaped metal staffs which are graduated in feet and inches to a sixteenth of an inch. Each end of the hose is fitted with control buttons which in use are pressed to allow air into the hose. The hose is filled with water to within a foot or so of each end.

In use, one staff is set up at the starting level point by one man while a second person selects a trial position a hose length of 15m away. Each one then presses his "atmosphere" button which allows air to enter and permits the water in the hose to find its own level. The water level is then read on each staff. People can become expert in the use of the Bunyip level with only a few minutes experience. Simple Surveying for Farmers by Professor Frank Debenham describes the full uses of the Bunyip level including its uses for do-it-yourself farm surveying and contour map production.

Levels of similar type, but which include a hose of opaque material, are some-times sold to or made by farmers and even recommended by some State departments. They should not be used, since with them it is impossible to be sure there are no air bubbles in the hose. At worst, the presence of large air bubbles makes any work done useless, and, at best, produces slow readings which are still of doubtful accuracy. If levelling work is necessary, then it must be accurate. A farmer should have sufficient knowledge and experience with the level he uses to be confident his work *is* accurate.

The irrigation channel may have a fall or be flat, according to the particular circumstances. For instance, in very flat country of 1 in 2000 to 1 in 10,000 slope, a contour and twin-bank irrigation channel will serve, while on land with a general average slope of 1 in 100, a single bank irrigation channel with a fall of only 1 in 5000 has been found to be steep enough.

A contour channel is often preferable. The disadvantage of a falling channel lies in the fact that too much of the flowing water could bypass the open water-gates and, after reaching the farthermost end of the channel, the water could build up sufficient height to overflow the bank. It becomes a necessity in such circumstances to place earth stops equipped with watergates across the channel itself to prevent that happening, and to direct the water into the irrigation bays. And each of the cross channel stops and watergates would need to be equipped with the same total width of water gateways as those servicing each irrigation bay.

Those measures may be unnecessary in irrigation channels built on the true contour, and those circumstances where cross channel stops would be an advantage, fewer stops would need to be placed, and at much longer intervals apart along the channel.

Considering such matters from all points of view, it becomes obvious that the lower the flow volume of water, the more necessary becomes the need for a fall in a channel and the more positive should the fall be as the flow volume decreases. Conversely, the greater the volume of flow and the more it is confined by banks and the deeper it becomes in the channel, the less is the need for a fall. But as Keyline flood-flow irrigation employs much larger flows than any other irrigation system there is no need here to consider lesser flows and their channel falls. There, a perfectly level or contour irrigation channel becomes the general type.

It may at first be thought that water will not flow in a channel on the true contour, and with small flows that may sometimes be correct. But when the big flow of water is discharged into the closed and banked-up end of a flood-flow irrigation channel there is nowhere the water can go, except to immediately build up depth and thus create its own positive down-slope and flow along the contour channel.

So, in levelling in of the irrigation channel, a contour line is customary and also, when there is any doubt as to whether or not a fall should be given to the channel, then a contour line should be pegged for observations to assist in making a decision.

Those who own an engineer's dumpy level will know how to use it. Those who do not should still preferably do their own levelling work, but with a water level as previously described.

Generally a true contour line should be pegged at each 15m interval. That distance is automatic and more than sufficiently accurate when the slack hose of the Bunyip level is used as a measure.

In pegging the line for the irrigation channel with a Bunyip level the instrument is first checked to see (1) that with the two staffs of the Bunyip standing upright and together the transparent hose is filled with water to within 0.3m of the ends of the hose in the staffs, (2) that there are no air bubbles in the length of the hose, (3) that when the atmosphere buttons or valves are both pressed and one staff lifted 152mm or so, the water in the hose moves freely or bounces, and (4) that the bouncing of the water ceases when the atmosphere buttons are again released. The level is then in working order.

One person places the foot of his staff against the starting peg on its upper side as the second person selects a trial spot for the next position by walking the full length of the slack hose, 15m away. Then with the two staffs upright, the atmosphere buttons are both pressed in and each person calls the water level reading on his staff. If the second person's reading is the greater, then his place is low and he changes his position in the direction of the rise of the country, again calling the staff readings until both are the same. If on the other hand the second staff reading is less than the other, the position is too high and is adjusted accordingly. In that manner the exact site for the second peg is determined and the peg is pushed into the ground against the staff on its lower side.

To fix the position of the third and all subsequent pegs, one of two movements may be followed. (1) Both people move together, number one to the position vacated by number two, where he places his staff in the exact position previously occupied by number two (staff against the peg and on the higher side) and number two moves to a point he selects as a trial for the position of the third peg which is then located as was the second peg.

The line is thus levelled in and pegged to some obvious limit such as the point of change from flat to undulating country, a definite stream or watercourse or a boundary, or to a suitable distance point such as 805m, 1.6 or more kilometres away.

It is to be particularly noted that the larger the flow volume of water under control, the greater the distance the water can flow efficiently in the irrigation channel. Moreover, as already stated, the cost of constructing the irrigation channel does not increase with the greater width necessary for an increased flow.

The form of the line for the irrigation channel

The line thus pegged may be fairly straight and even, which will indicate that the form of the land is large. The line may be in the shape of one large uniform curve which indicates, in the case of the curve being toward the rise of the land, a large but very flat valley formation. When the curve is down the land a large flat ridge formation is indicated. Again the line may have uniform curves swinging alternately towards the rise of the country and away from it which indicates alternate flat valley and ridge formations. Then the pegged line may be in any of these forms but be far from smooth or uniform. As a channel of smooth and regular line is desirable for construction and for the practical management of the irrigation area, the last-mentioned line is adjusted or normalised. That is done by walking the line and examining the out-of-line peg positions for an alternative site which would improve the line form.

It must be remembered that the present shape and form of most of the flatter land has been stabilised as a result of its vegetation, rainfall, winds and flowing water of past ages. While those land shapes then do have some variety, they all tend to produce smooth curves in their contour lines. Where a badly shaped line results from the contour level pegging it is often the result of some form of soil deterioration. For instance, vast areas of flat land which have suffered from recent wind erosion caused by unsuitable stock management practices or a series of dryer than average seasons with high winds. Those conditions may have produced local variations in surface levels of several centimetres in a distance of only 15m, in circumstances where the general fall of the land may be only 305mm in 1.6km. In such awkward conditions those with the level should try to follow the remaining spots or areas of original surface or, if that is impracticable, they should avoid all noticeably high and low spots. The high spot to avoid will usually be associated with scrub plants where windblown dust has been deposited, and the low spot between bushes. The greater the extension in length of the contour line from the starting point, the more the true contour of the original surface discloses itself, and that is the line required for the construction of the banks for the irrigation channel.

So in walking the pegged line to adjust it, it is well to carry pegs of a distinctive color. Then, when an original peg appears out of the general pattern of the line, the spot is examined to see whether because it is on a local high or low spot, it can be moved up the land or down the land and so improve the line. The old peg can be left for the moment and a new peg of different color placed in a better line position. A better and smoother shaped line can thus be produced and still remain for all practical purposes on the true contour. We have called the action of adjusting the line "normalising".

Although it is important that the general aspect of the line is that of a contour, it has been found from experience that purely local small variations in level matter little and even less so as larger flows are used. Nevertheless on land generally uniform, it has also been found that a small adjustment to the position of a few individual pegs will always improve the uniformity of the curves and thus enhance the overall appearance of the final channel. "Normalising" of a line does not aim to produce a straight line for an irrigation channel where the contours indicate curving forms.

However, no adjustment should be made to improve the line during the actual levelling-in of the line, other than by avoiding the local high and low spots. Levelling and pegging should always be completed first so the true line is more readily disclosed by the pegs which have been accurately placed with the levelling instrument.

If the irrigation channel is of the single bank type, the levelling line completes the pegging for the irrigation channel. If the channel is of the twin-bank form then the second bank is pegged by measurement only above the first line of pegs. From each peg of that line the set distance is measured at right angles above it and the new pegs are placed through to the end of the line. The distance apart of the two banks is determined by the factors already mentioned. In general, the channel should carry its full flow of water with less than 305mm depth of water against the single bank or against the lower of the twin banks.

Constructing the irrigation channel

It is an advantage to have a clean line with which to work. Unless the grass is short it should be cleanly mowed for up to 245mm above and below the pegged line and the cut grass moved away from the line.

A strip of land 2.4m wide extending from the levelled-in and pegged lines on both sides is now cultivated two or three times with a chisel plough. The course of the cultivation should match-in with the pegs in such a way that smooth curves are formed, and not a series of short straights from peg to peg. The objects of the cultivation are twofold: First to ensure that the new earth bank will bond into the earth below it, and second, by cultivating each time with the chisels in the furrows of the first run, there are provided guidelines to help the work of the sidecasting implement forming each bank. If possible the bank should be constructed while there is obvious moisture present in the earth. (See cross sections of single and twin bank channels figure 15, above.)

The first sidecasting run is made with the lower and rear point of the blade of an angle bulldozer or grader digging 0.9m from the pegs and throwing the earth away from the tractor and toward the pegs. The pegs may then be recovered for later use. The earth placed by the first run is consolidated by run-

Cross section of a single bank (A) and a twin bank (B) irrigation channel for Flood-Flow. No fall.

ning the tractor along the line with one track of the tractor on top of the loosely placed moist earth bank. The second sidecasting run is similarly done but on the opposite side of the line, and again throwing the earth on to the pegged line.

Each sidecasting run should be continuous for the full length of the line. The sidecasting implement should be driven at a good fast walking speed so as to obtain a suitable "throw" effect to the earth. While low gear of a crawler tractor is too slow, second gear at full throttle is usually effective. A tandem drive grader or an angle-and-tilt bulldozer of medium horsepower are the most suitable machines for the construction of the banks of the irrigation channel.

After every sidecasting run, the earth of the bank is consolidated by running the tractor along the newly placed earth of the bank with one track on the top of the bank. Two or four sidecasting runs are needed with that type of equipment throwing the earth from each side alternately. When completed a full-sized bank should be about 1.8m wide at the base and 0.6m high.

During the construction of the bank the earth is taken from immediately along each side, which leaves depressions in the form of excavated drains. The drains may be stopped with small earth banks 15m apart. They will be gradually filled and levelled off in the management and improvement of all the land of the irrigation area. The cultivation, sowing, eatingout and occasional mowing and harrowing of the irrigated land includes, of course, the land of the irrigation channel which soon is fully grassed and stabilised.

In conditions of climate that are hot and windy, it is desirable to adopt special measures to quickly grass the irrigation channel banks. By placing earth stops in the excavated drains beside the bank, ponds of water will lie along the sides of the bank after each irrigation, and the ponds, by keeping the adjacent soil moist for some time after each irrigation, will greatly assist the spread of grass over the bank.

In dry conditions and with the bank of the irrigation channel completed, water if available, may be released into the channel. The first flow of water may serve as a partial check on the work and it also will provide suitable soil moisture conditions for the later completion of the work. If the banks must be completed with the soil very dry then it may be better not to attempt to consolidate them but await natural settlement which will occur after rain has fallen.

Watergates

The construction of the bank or banks which form the irrigation channel is completed as described, namely in long fairly fast sidecasting runs which produce uniform banks. Now the banks are to be broken at intervals to provide outlets for the irrigation water into the individual irrigation bays. It is done by the installation of watergates in the single bank channel and in the lower bank of the twin bank channel.

It has been found that the most practical size for the watergates is 1.8m wide by 0.6m high. There are two parts to a watergate: (1) a frame in channel iron form which has a continuous fin projecting outward from the bottom and both ends and (2) a separate closing panel formed of sheet steel. The closing panel slides in the channel form of the gate frame and an effective water seal is obtained with a strip of sponge rubber which is attached around the contact edges of the closing panel.

The openings in the channel bank for the watergates can be made by using either construction equipment, a farm tractor-attached blade or by hand shovel, but in any case the work will need to be finished off by hand. The earth from the openings made in the bank is placed in heaps on the undisturbed bank at each end of the water gateway. The bottom and end fins of the frame are let into the earth by digging a narrow trench to take the fins. If the earth is moist and soft it is often possible to set the gates without actually excavating the small trench. On one such occasion an old meat cleaver was hammered into the earth to form a continuous slit of appropriate length and depth and the watergate fins were then tapped into the slit. After placing the watergates the earth should be rammed into a tight contact with the fins and on both sides. A 25mm round or square rod is suitable as a rammer.

Whatever method is used to place the watergates it is essential that (1) they be placed in sequence from the start of the irrigation channel, (2) each water-gate as placed be levelled in from the previously placed gate, starting from the supply end of the channel. Where the channel is dead level the sills of all water-gates will likewise follow the same level, and where the channel has a fall, each successive watergate will have the same rate of fall according to the distance apart of the gates. (3) The sill of the first gate must be on the same level or a little lower than that of the datum of the channel, (4) the sills must be perfectly level, (5) the earth around the fins be well rammed into a tight contact, and placed as tidy heaps on the bank at each end of the openings.

It has been determined from experience that one such watergate will release irrigation water at up to the rate of $2250m^3$ (half a million gallons) an hour and not produce any difficulties for one-man control or cause any washing away of soil.

If a flow of water of much over $2250m^3$ an hour has only one watergate outlet, then the level of the flowing water in the channel will rise to such a height that the increased depth of flow through the watergate will eventually accommodate the full flow. There is a limit to the depth of flow to that of the height of the closed watergate. Maximum of the closing control of the standard watergate is 508mm.

In practice, a water level of 305mm at the watergate should be considered as the crest which is both practical and economic. A greater flow from a watergate can produce two problems. (1) It increases water flow velocity on the soil adjacent to the gate and immediately inside the irrigation bay; this is the one critical velocity point in the whole of the layout for the Keyline flood-flow irrigation system. (2) It imposes an added depth of water against the irrigation channel bank. It should be noted that the technique of sidecasting for the laying in of the earth for the irrigation channel bank, while being very low cost, has definite limits in the height to which the earth can be placed with an extra few millimetres of height to the bank involving a disproportionate amount of work.

Although the suggested constructed height of the bank is 0.6m, the bank will always flatten somewhat during the use and management of the irrigation area, and so the banks lose some of their effective height. However, the chief consideration must remain one of velocity of flow at the watergate entrance to the irrigation bay. Consider these facts which relate to velocity of flow at the watergate. The width of the irrigation channel is such that even when recently chiselled, the velocity of flow in the channel with its slight or no fall is so slow that wash does not occur. Likewise, the width of irrigation bays may vary from 2½ to five or more times the width of the irrigation channel so that, even with a fall, no wash is likely to occur. But for example, where an irrigation channel 18m wide supplies irrigation water at a rate of $4500m^3$ (one million gallons) an hour to each irrigation bay, which is 45m or more wide, it may do so through two watergates with a combined effective width of a little under 3.6m.

The increased velocity of flow which occurs in each bay immediately at the water's entrance may not have any detrimental effect on a sward of pasture established at that point, but, as the water flows over some recently disturbed earth during the first irrigations of a new area, definite proceedings to stabilise the points in each irrigation bay may be at times necessary.

It has been found that the most practical procedure is to start the first full irrigation as soon as possible and use the watergates with a full flow of water. One or two men with shovels should constantly patrol the irrigation channel area to adjust any low spots or leaks in banks, which may occur. Every effort should be made to maintain the maximum flow of water so that a full test of all work can be completed. Pegs can be placed if needed to identify particular places for later adjustment. All watergates should be used to their full capacity without any regard to possible wash. Moreover, the water flowing through the watergates may be cloudy on the first use and any wash which may be occurring will not be immediately visible through the water.

On the day following the first irrigation test, the areas near all watergates are inspected and where wash spots are found they are repaired using earth previously heaped on the bank at the end of each watergate. The repaired spots should be planted with grass seeds, and fertilised if appropriate, and firmed down by stamping. As all watergates are opened the day following the completion of each irrigation those spots are therefore constantly under observation.

The cost of the watergates on a $4500m^3$ an-hour project may be $30-40/ha of irrigated land. On a particular project the cost may provide for irrigation bays with an average area of 4ha, with each bay being serviced by two watergates. The time of flow to each bay would be about 30 minutes and the average water intake by the soil under the conditions of good management should be about $500m^3$ (50,000 gallons), or a little over 51mm of water a hectare at each irrigation. If both the flow rate and the bay size are doubled, the cost of the watergates a hectare remains the same, although the labor cost, now extremely low, is halved.

In conditions of extremely unstable soil, an extra watergate can be installed temporarily to each bay and taken out after the soil has improved and become more stable, and the gate used for later irrigation developments.

Watergates for a purpose other than the release of irrigation water will be required on some flood-flow irrigation projects where the twin banks irrigation channel are in use. In circumstances where rainfall runoff would pond against the upper bank, watergates should be installed and opened after each irrigation so the runoff will move into the irrigation channel. Then with the watergates open from the irrigation channel into all bays, the rainfall runoff is distributed uniformly over the whole of the area. The comparative number of gates in the upper channel to those in the lower channel is not determinable for all rainfall conditions, moreover the volume of runoff will also depend on the area above the channel which will produce the flow. One watergate in the upper bank to each four in the lower bank is therefore given as a general recommendation only.

From that description of the design and construction of the channels for flood-flow irrigation on flatter lands, it will now be apparent that the specially designed channel has outstanding advantages over the orthodox flat land water channel. First, there is the more economical use of earth with the attendant lower cost because the water in the usual channel flows principally in the earth which has been excavated for the purpose. The water flows on the surface of the land in the flood-flow channel and only sufficient earth to hold it there needs to be used. For an equally large flow of water its cost could be less than 1/10th of the cost of the orthodox channel. Second, because the water flows on and above the surface of the land it is most suitably placed for release into the large irrigation bays by the simplest and lowest costing means. Third, the relatively small banks do not require crossover bridges and do not constitute any obstruction to stock movement. The passage of vehicles across the small banks is a simple matter of flattening the banks at suitable places.

There are other advantages in management, as the area of the flood-flow irrigation channel itself is far from wasted, but instead is part of the irrigation land. It may be sown to pasture and grazed, harrowed and mowed with the rest of the area.

It is worth emphasising that the total land preparation costs for flood-flow irrigation (including the formation of the irrigation bays by the water steering banks, which are discussed in the next chapter) are even less a hectare than the low cost of the watergates.

CHAPTER ELEVEN

Flood-flow steering banks

IRRIGATION BAYS FALL FROM the irrigation channel down the maximum fall of the land, and all watergates are opened the day following irrigation and remain open until the start of the next irrigation. Those two factors, in conjunction with the correct width of both the irrigation channel and the irrigation bays in relation to the volume of the irrigation stream, provide in advance for some of the major requirements for the good water management of irrigated land. The requirements of management involve control of the soil's water intake so as to prevent overwatering. That is achieved in the design which balances the width and the length of the irrigation bays with the volume of flow so that the irrigation water remains in contact with the soil of the bays for only half an hour.

On all soils on which Keyline principles of management have been applied, the length of time of water contact with the soil can be halved to 15 minutes with no disadvantage. Even an occasional doubling of the time to one hour will not have any substantially detrimental effect, except to waste water, particularly when the soil is light and sandy and the project is new. It should be noted that the smaller the irrigations stream in relation to the bay size, the longer is the water in contact with the soil and the greater is the chance of overwatering and waterlogging. Destruction of soil by excess water and poor drainage has been, unfortunately, part of the history of irrigation for thousands of years. Therefore, it becomes necessary to make provision for good drainage. It is ensured in the first place by the increasing fertility of the soil produced by Keyline soil management techniques and, in the second place, by the fact that the steering banks which form the sides of the irrigation bays follow the steepest path down the land and therefore the most efficient surface drainage lines.

The Keyline flood-flow layout of the irrigation channel, of the steering banks and of the associated bays, because of their combined but confined spreading effect, gain the greatest benefit from rainfall and at the same time provide the best surface drainage pattern. In many circumstances the whole layout would still be worthwhile employing on land which could not be irrigated, solely for the benefits to be gained from the control and even spreading of the water from runoff rainfall.

It is considered of particular importance that the design and layout of land for any type of irrigation should be such that the water from occasional heavy and continuing rain immediately after an irrigation is handled effectively and automatically and without damage to the soil. Hence the importance of having watergates remain open after each irrigation is completed when there is any chance of rain. Stated briefly, "good irrigation layout is automatically good drainage layout".

However, those critically important factors are only rarely achieved by haphazard methods. For a start the naked eye cannot pinpoint the maximum fall of the land on relatively flat country. In fact the eye cannot with certainty determine uphill from downhill on land of only a moderate degree of flatness if there is no stream or watercourse visible to guide it.

The maximum fall of the land is one of these extremely important "water lines" of land, and a levelling instrument is needed to determine it or to at least "guide the eye". But there is another point to mention before proceeding. Eventually there will be more intimate subdivision of the irrigation land into suitably sized paddocks, and some fencing will be required other than that which divides the new irrigation land from the adjacent rainpasture and crop land. Of necessity, one or more fences will cross the irrigation land downward from the irrigation channel, so it is as well to consider the general pattern of subdivision fencing at the time the positions of the steering banks are being determined.

If, for instance, a million gallons of water an hour flows into an irrigation channel 2000m long and supplies an area 2000m deep, this area of 200ha will certainly need dividing in two and subsequently into four and more paddocks. The subdivision fences should not "fight the water" by crossing the bays on an angle, but should obviously "go with the water". Fencing is therefore associated with the pattern of the steering banks. They should follow the steering bank which, first of all, suitably divides the area in two. It is a simple matter to divide the area into two roughly equal paddocks, but even so a starting point for the fence must first be selected.

There may not be any critical factor relating to the selection of a starting point for the steering bank and its associated fence if the starting point is along the irrigation channel. But there could be such a situation as to where exactly a fence should finish at the lower boundary of the irrigation land. That boundary is sometimes a stream which may be suitable for stock watering. By locating the starting point of the first steering bank and its associated subdivision fence at the lower boundary, a selection may be possible which would provide water for the both paddocks instead of only one of them.

The finishing point in the locating of a steering bank is not at all obvious from a selected starting point, and so a starting point, if selected on the irrigation channel boundary, would not "come out" at the desired point for say, stock water facilities which may exist on the lower boundary of the irrigation area.

Steering banks are located by a levelling instrument — not by eye. If there are no such considerations as above to effect the decision on a starting point, it may not matter whether the first steering bank is levelled-in from the upper or from the lower boundary. Levelling-in upwards and levelling-in downwards on flatter lands are in the nature of reverse proceedings, but it is not always so when there are even slight undulations to the land. For instance, to find the exact position of the steepest path in a valley the levelling must proceed downward, whereas to find the exact centre fall of a ridge shape the levelling must proceed upward.

Levelling for a steering bank

If the selected starting point is on the lower boundary, the procedure, first with a Bunyip level and second with a dumpy level is as follows: A peg is placed at the selected starting point for the first steering bank, which may be one which divides the area into two nearly equal parts, and one staff of the Bunyip level is set up against that peg on its upper side. The second man then places his staff 15m away, the distance which is governed by the slack of the transparent hose of the Bunyip level, and on what he considers to be the direct uphill direction. The atmosphere buttons are pressed and readings are taken by each man and the variation in the height of the two staffs noted. The reading of the forward staff is particularly noted. Now, with the back man maintaining his staff position, the forward man begins to take more readings as he moves his staff to three or four positions on an arc of a circle, with the radius of the circle represented by the 15m measure of the level. He selects the point, the lowest of the series of readings, which is then the highest spot 15m distant upland from the start, and places a peg against his staff on its lower side. Now both men move forward the back man to the second peg just placed by the forward man, and the forward to a new position as before, where by taking readings on an arc he again finds the highest point and then places the third peg.

That method of pegging continues upwards to the irrigation channel and finishes about 20m beyond the channel on its upper side. A longer peg or a steel post for sighting may be placed at the last point.

If the vertical height from the start to the end of the line is required, then a record of each set of readings of the Bunyip is kept. The total of the differences in readings then is the vertical height.

Dumpy level

Three people work when levelling and pegging the line of a steering bank. They are an instrument man or surveyor for the level, a staffman and a chainman.

The dumpy level is set up on a location at a suitable distance up the land from the first peg to enable the surveyor to take reasonable advantage of the capacity of the telescope and thus be able to read a number of sitings from each setup of the level. The staffman positions his staff at the first peg and the surveyor records the first staff reading. The chainman remains at the first peg holding one end of the measure while the staffman, holding the other end of the measure, walks to and holds his staff upright at a spot he selects as being on the line of maximum rise. The measure may be a surveyor's tape, a length of light rope or wire and may be 15m, 20m, or even 30m long if the land form is large and smooth. The surveyor notes the new reading on the staff and then signals the staffman to new positions both to left and right on the arc of the measure, determines the high point and signals the staffman to place the second peg. The chainman walks to the newly placed peg as the staffman, with his end of the measure, walks to a new spot upland as before.

180

One end of the measure can be looped around the staff, and each reading of the staff should be taken with the measure fully extended from the chainman to the staffman.

As previously indicated, the levelling-in of a line for a steering bank may usually be done on very flat land either from the lower side of the irrigation area and working upward, or from the irrigation channel and working downwards. In the latter case, each sighting locates the lowest point on the arc of the measure and thus the greatest reading on whichever of the two types of levels is used.

The line of pegs thus positioned represents, when viewed from the sight peg placed near the irrigation channel, a local line of maximum fall directly down the land. As previously stated, this is an important and distinctive water-line-of-the land and if a series of the lines was plotted and marked in for various properties in such a way that would enable them to be studied, it would be found they would have produced various patterns which are as distinctly individual for each area as the contour lines themselves are for different areas. Further, it would probably be found that the patterns produced could be classified or grouped into definite categories or classes, as are the patterns of contours of land grouped and classified in Keyline. Also it can be realised that if a very large stream of water were discharged continuously at the topmost peg just below the irrigation channel, then the water would flow down the land on the line of the pegs, albeit spreading wider as the particular land is flatter.

That line of pegs is now considered as the general line for a subdivision fence. While it may be quite different from what may have been expected, it will, however, disclose the true fall of the land, so that if it is now found to be unsuitable for the fence line it will at least have indicated where the fence should be located.

Closer examination of the line of pegs may show that it is in one of four general patterns: (1) It produces two curves in the form of a flat S bend, (2) one large curve or arc to the right, (3) a similar curve but swinging in the opposite direction, or (4) a line generally straight. Also, as with the pegging of the irrigation channel and for similar reason, the line may be irregular and need adjusting.

The closer to the first line that a second line is pegged in that manner the more closely generally will the two lines parallel each other down the slope. And once a few lines have been marked in it may become apparent that it is unnecessary for the lines for every steering bank to be located so precisely with the level. However, whenever there is any doubt about the course to follow the level alone should guide the issue.

In the conditions previously mentioned, 45m is a suitable distance from one steering bank to the next. Thus, if the first line does not appear to divide the area satisfactorily, a second line is levelled-in for the fence line. Once the two lines are pegged in, the rectitude of these proceedings becomes so obvious that even a totally inexperienced person is immediately convinced there is just no other way for the irrigation water to flow except downhill, and that he has, with his levelling and pegging, achieved a perfect and scientific "downhill" line.

When a steering bank line is to be also a subdivision fence line, it is of considerable advantage if the line be "formalised" into straights more suitable for fencing. As the "formalising" to straights of the curves of the various water lines, including the present maximum slope lines, contour lines and the lines of falling channels is somewhat of an art as well as a science, it will be dealt with more fully in the next chapter.

The line for the steering bank under discussion is in the middle of an area which may represent the first stage of a particular flood-flow irrigation project, which is itself only a part of the Keyline plan for a farm or a station. The steering bank lines should therefore be located from the middle section back to the start of the area where the irrigation water flow starts; That section could be completed quickly, and thoroughly tested with a full flow of water.

If the first two steering bank lines indicate a definite uniformity of land slope, the third line to be levelled-in can start at a point twice the distance away as the width of the irrigation bays (in this case 90m). Then with the third line levelled-in, a fourth line can be located by measurement half way between the second and third line and pegged about 90m apart. As soon as that is done, it will be again apparent whether that method of locating only alternate steering bank lines with the level is suitable for the particular circumstances of the area.

Constructing the steering banks

The implements most suitable and economic for the work include a 2.1m or wider chisel plough and a small to medium sized angle-and-tilt bulldozer. The ordinary farm tractor and rear blade ditcher is usually quite suitable in place of the small angle bulldozer, requiring only that more cultivation be done before-hand and that a greater number of sidecasting runs be made. It should be under-stood that a farm tractor and ditcher is a very unsatisfactory combination for digging undisturbed earth, but very satisfactory for moving earth already dug — well cultivated.

The contruction procedure follows that employed on the irrigation channel bank, except that the steering banks are only half the height and contain less than one-quarter the amount of earth. Again, unless the grass is short it should be closely mowed and the grass moved off the line.

The path of the first chisel run is made with the centre of the plough over the top of the pegs. The pegs are then recovered for later use. The second and third runs with the chisel plough are made in the exact furrows of the first run so that the strip as finally cultivated is still only 2.1 to 2.4m wide.

The small angle bulldozer makes only one run from each side with the digging point of the blade in the second chisel furrow from the outside of the cultivated strip, and thus sidecasting the earth towards the centre of the strip. After the first sidecasting run the newly placed earth is consolidated and bonded in by travelling the full length of the new bank with one track of the tractor on the top of the loose sidecast earth. The second run is then made from the same position but on the opposite side of the cultivated strip, and again the new earth may be consolidated with one track of the tractor. However, if the work has to be done when the earth is extremely dry it may be better not to try to roll in the earth with the tractor.

In conditions where a grassed area is sod-bound the above procedure, while producing a bank which works satisfactorily by steering the irrigation water and not leaking, may form a bank which is very lumpy and somewhat unsightly. In those circumstances, the number of chisel cultivations can be increased with the first two or three runs made with the chisels spaced closer together, and with short weed knives attached, but finishing the final run with the standard 305mm chisel spacings.

In these same conditions, a first shallow cultivation run with a rotary hoe to chop the sod, when followed by the two or three chisel cultivations, may assist in producing steering banks of smooth finish.

A first shallow cultivation with a disc or mouldboard plough, if available, may be of assistance in a similar manner, but those implements are generally unsuitable for any other purpose on irrigation land. Their inevitable and continual sidecasting of soil at every run tends to change the natural "water shapes" of land, thus interfering with the uniform spreading of the irrigation water.

The lower ends of the irrigation bays, of which the steering banks form the sides, are left open so that runoff from heavy rainfall and drainage water can get away. The runoff water from the irrigation bays may flow directly to a stream below the area or continue on to lower land which itself eventually drains to a stream somewhere below.

The upper ends of the steering banks, as formed by the equipment, do not join in properly to the bank of the irrigation channel and some hand shovel work is necessary to complete and tidy the steering banks at the junctions.

Further, the construction of the steering bank in the above manner leaves a small drain on each side of each bank, throughout its length. Therefore earth stops are placed in those drains 10-15m apart along their full length. The earth for the stops is taken from high spots on the steering banks and high places within the irrigation bays. The earth stops continuously turn the main flow of the irrigation water back towards the centres of the bays. They should be higher than the level of the land and they will need checking and attention for the first few irrigations to see they perform their function. Those stops are of great importance to fast economical irrigation. They also cause small ponds of water to lie against the steering banks after each of the first few irrigations. The ponds could greatly assist the grassing over of the raw earth of the new steering banks. As with somewhat similar circumstances associated with the banks of the irrigation channel, time and good management effect the virtual disappearance of the small excavated drains which are associated with the construction of the steering banks.

CHAPTER TWELVE

Fencing on the water line

NORMALISING, IN THE SENSE USED PREVIOUSLY, is the art and science of marking in on deteriorated land with slightly altered surface, practical and workable contour lines and falling lines for channel, drains and guidelines for the control of cultivation patterns.

As mentioned previously, recent wind erosion on some lands of little slope will cause disparities in the surface of land and produce variations in the local readings of levelling instruments, which may be as great as the general difference in levels found in a distance of over 1.6km. Thus a local variation, at only 15m distance, may be equivalent to that of the general fall at a distance of 305m or more away. The object of normalising then is to produce the lines as the contour would have been if some of the surface had not been slightly altered by wind erosions.

Formalising on the other hand is the art of producing from the smooth shaped contour curves themselves of normal land forms, straight lines to replace the curved lines and, at the same time, lines which maintain the principal values and virtues of the true contours.

The arts of normalising and formalising are also applied to those other water lines referred to in the discussion on the planning and the levelling-in of the steering banks.

The function of formalising, as it applies to the fencing of irrigation land, has two forms. One may on occasions apply to the fence along the irrigation channel or similar contour divisions, and the other to those interior subdivision fences of irrigated land which go down the fall of the land. Both forms also have an important application in the subdivision of non-irrigated land.

Dealing firstly with the upland-downland type of fence which is to be placed on a steering bank, and referring to figure 16 opposite, the series of crosses represents the pegs as placed with the aid of a levelling instrument and according to the instructions given elsewhere. The wavy line represents the line as normalised; the line formed by the series of straights then is the formalised line. Again the series of crosses represents the true line as pegged with the aid of the levelling instrument; the wavy line, while being off true, represents the best curved form of the pegged line which is practical for the construction of the steering bank; and the line form of the series of straights represents the nearest to the true line practical for a steering bank which may have a fence on it.

In some circumstances the pegs, as represented by the crosses in figure 16 may not favor normalising because of wide variations from any form of line. They will, however clearly indicate the true general downhill direction and so one or more straights may be more suitable when selected directly than by first normalising the line.

*   *   *   *

The formalising of the irrigation channel line for a fence position aims to produce the "straights" either above or below the appropriate bank of the channel

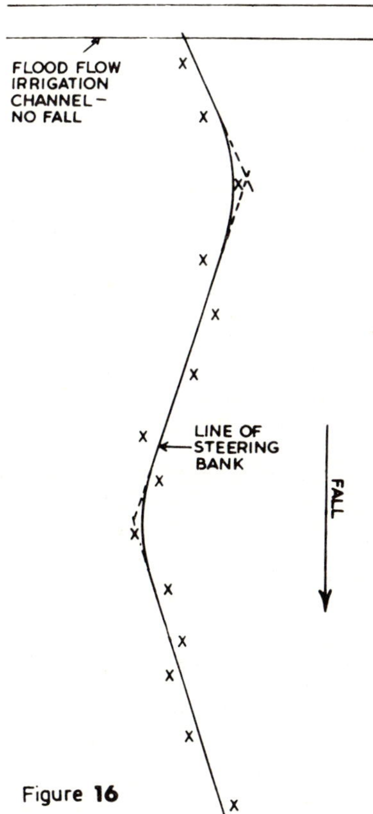

Pegging the line for a steering bank. The crosses are the points of maximum fall located by the level. The curved line represents the steering bank. The three straights show the formalised line of the steering bank if a fence is to be located along it.

Figure 16

and parallel to it in a general way and at a specified mimimum and maximum distance from the bank of the channel.

For beef and other types of general production it has been found from experience in the management of irrigation land, and including the various systems of flood and flow, that where water is distributed from channels, 6.4m and 9m are suitable minimum and maximum distances for the fence position from the nearest bank of the channel.

Referring to figure 17, the wavy line represents the upper bank of a twin-bank irrigation channel in which the irrigation water flows from left to right. The irrigation land lies below the line and the fence is planned for above the top bank. The fence is to divide the irrigated pasture from the rainpasture land above it.

The work, as with all other planned land development, may be first plotted on paper if the area has been suitably mapped. The result can thus be "seen" in a general but practical way. The work will also be measured and sighted-in on the land itself, and for that the procedure is as follows: One man can do on his own the pegging of the fence line but a helper is a big advantage. Ten pegs 1.2 to 1.5m and white-topped for easy sighting will be needed. A chain tape may be carried, or the surveyor may rely on "pacing". The term surveyor is here used in the wide sense for the man who makes the decision, as well as in the narrow sense; he may then be the owner of the property, the manager, a supervisor, a foreman or the boss for the time being.

Figure |17|

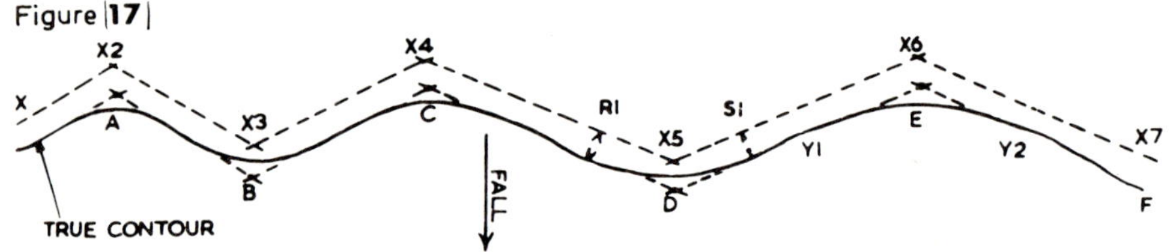

Locating a fence along an irrigation channel to divide rain land from irrgation land.

Referring again to figure 17, which represents an existing channel, the first peg is located at X by measuring 6.4m north of the bank and at a right angle to it on the left or west end of the bank. The surveyor then proceeds to point A, which he locates by sighting to right and to left along the two curves and then picking the point A as would be the junction of the two lines if they were extended until they crossed. He places a temporary peg at A and then measures 6.4m up from A and directly away from both straights. That spot is pegged as the second fence point, X^2. Sighting back from the fence point X^2 to fence point X, he will see that the straight of the fence will be the required 6.4m minimum distance everywhere from the bank and that the maximum distance from bank to fence is at X^2.

He now proceeds to point B which he locates precisely as with point A, only this time the point is below the upper bank and in the channel itself. He now walks north on a line which divides in half the angle A-B-C and measures the 6.4m minimum distance, but this time the starting point for the measured distance is from the bank itself as at X. The third peg of the fence line is placed at X^3.

In review, he has now placed X the starting peg; X^2, a peg in correct location in relation to the flat valley shape which the curve upland represents; and X^3, a peg in correct location in relation to the flat ridge shape which the curve downland has denoted.

He proceeds in like manner to C valley shape and places X^4, on to D, ridge shape, and places X^5; now to E a larger and flatter valley shape and places X^6; and so finally to F where X^7 is placed in the same manner as the first peg X.

That is the completion of what could be the final line of the fence, or it may be altered after checking. The surveyor now proceeds to examine the fence line by checking the places where the line reaches the maximum distances from the channel bank. The obvious points to examine are R, S and at peg X^6.

The only ways to retain the minimum distance while reducing the maximum distance are (1) by making an extra straight in the fence line, and (2) by the inclusion of a gateway at one of the pegs which now represents a corner of the fence line.

Referring to R and S on the diagram, at which points the fence is farthest from the channel bank, new fence points may be located at each in the same manner that fence point X^3 was located from B. An extra straight, R^1-S^1, would result in the fence to bank distance being equal to the minimum at points R and S. In this task, care is taken to see that some other points on the lines X^4 to R^1 and S^1 to X^6 do not now fall within the minimum distance of 6.4m, and thus require some little adjustment upland of points X^4 and X^6. It thus becomes apparent that the addition of a new straight to what can now be regarded as the first trial line may make it necessary to check the fence points again from start to finish.

Gates, located at fence points, by being in effect short straights, act in a similar though lesser manner to the extra straight R^1-S^1. Before the final fence line is decided upon gateways need to be located. And in the matter of gates on irrigated land, two general errors are to be avoided. First, too few gateways are usually provided and second, the gates themselves are not good enough for the job.

Referring to the case illustrated in figure 17, there are at first sight four obvious gateway positions, namely at X, at X^3, at X^5 and X^7 and they should all be used. Even so, the distance apart of gateways at X^5 and X^7 is too great and at least one other is required. The selection of a site for that gateway is at one or the other of points Y^1 and Y^2.

Now the reasons for the selected gateway sites are, respectively: X is closest to the delivery point of the water and a gateway for service and management is required, and here it may also serve for stock movements. X^3 is on the centre line of a ridge and also lies at a suitable distance from gateway X. Points X^2 and X^4 are not so suitable since they lie in the centres of flat valleys, and valleys do not usually withstand traffic and tramping as well as ridge shapes do. X^5 again is on the centre line of a flat ridge and its position is suitable in relation to the gateways at X and X^3. X^7 at the limit of the channel is another "must". Now only a gate between it and X^5 needs locating.

Point X^6 is not desirable since, like points X^2 and X^4, it lies in the centre of a valley form. The decision as to which of the points Y^1 and Y^2 is the one most suitable for a gateway can be decided by reference to the general direction of the steering banks in that part of the irrigation area. For instance, if the steering banks swing to the west in relation to the eastern boundary fence which may run north and south, then a gate at Y^2 would be the one selected, as the fence line downland which would be associated with that gateway will then divide the area more equally. If, on the other hand, the steering banks tend to the west, so for similar reasons, Y^1 would be the chosen site of the selected gateway.

Now with the line of the fence clearly marked with the corner pegs and the position and exact width of gateways pegged also, the surveyor walks the line for the final adjustment of the fence position. The fence line now comprises a series of longer straights which represent the fence itself and a series of short straights which indicate the gateways. Final adjustments of the whole fence line then are designed to maintain the fence at not under the minimum distance above the top bank of the irrigation channel, but at the same time to keep down the maximum distances from fence to bank.

The first time a fence of that type is located by a surveyor he would do well to follow the foregoing procedure in detail. However, once he is experienced by having completed his first such fence, he will be able to shorten the work by deciding on and pegging the gateways and the complete fence line at the one time.

The procedures for positioning a fence above a single bank irrigation channel are exactly the same as those just described except in one respect only — the minimum distance has to take into account for that type of channel the water flows on the land above the channel. Either the average width or the maximum width of the water flow is added to the 6.4m which was used as the minimum width for working and management. So if the stream flows 11m wide above the channel bank then the figure for the minimum distance above the bank for the fence line is 6.4 plus 11, or 17.4m.

An irrigation area needs its fair quota of trees, for stock shelter and for windbreaks, and the belt or lines of trees can be suitably associated with the subdivision fence which divides the irrigation land from the rainpasture land above it. For two rows of trees of medium spread, 14m will need to be added to the minimum distance of 6.4m so the minimum distance would be 20m above the bank, or above the water line as the case may be, for the permanent fence line. A second fence below the permanent fence is then a necessity for the protection of the trees from stock during the first three to five years of growth of the trees. The second fence can be classed as a temporary fence but it must still be fully stockproof to serve its purpose of protecting the young trees.

Farm and grazing boundary and subdivision fencing needs will never be served by only one general type of fence, and only by classifying fences into types or categories will logical decisions on the selection of the right fences for various purposes be possible.

The three main categories of fences are (1) boundary, (2) subdivision, and (3) stock handling, and while the differences of the three categories are fairly wide, there are other wide variations within the categories themselves.

The boundary fence is the only type of fence which is precisely located, and its straights and corners permanently fixed, without the necessity for any decisions on the part of the owner. The construction means and materials used invariably follow local, district and sometimes national usages. In general, the boundary fence fashions of the district can be followed with some certainty that they are likely to be the most suitable and economic under the local circumstances.

The introduction of fence wire and wire netting to counter the multiplication of the rabbit population completely changed the fashion of boundary fences. But today with the rabbit no longer considered a menace in former infested areas, it is becoming less unusual to encounter boundary fences built without regard to the former pest of the rural areas. The new fashion in boundary fences which may now be emerging appears to be based on the assumption that the rabbit can be kept permanently under control. The dingo or wild dog is still troublesome in some areas and they influence the design of boundary fences.

Subdivision fencing is, however, the principal type which now needs greatly improved technique of planning and location, as well as improved methods and material in the building of the fences themselves.

Adequate subdivision fencing is a major capital cost on any rural landholding, and on many occasions its final cost will be much higher than the purchase price of the farm or the station. But fencing which subdivides land to the best advantage also permits the manner of control which is itself a major factor in soil improvement as well as in production. Holding stock on the pasture of one paddock for too long a period is somewhat like trying to produce sugar by cutting the top off the cane each time it appears above the ground.

Good grass, like all other crops, needs time to grow, when all it wants is letting alone so that a good crop can be harvested by the stock later on. And if the grass is harvested at the right time by the stock, good pasture grown on fertile soil is still by far the cheapest and best stock food of all.

But there is another great benefit from adequate fencing which permits of management which gives the grass of each paddock its time-to-grow. When grass grows upward its roots grow downwards and deeper as the grass approaches the flowering stage. And since the best and lowest costing organic material is the recently dead roots of a good pasture, and that organic material is the basic food of soil life and which produces the climaxes in soil development , the deeper the roots and the greater the quantity of root material produced, the better for the whole dynamics of soil life. So with the aid principally of good and sufficient fencing and subdivision paddocks, pastures can be controlled and managed to produce more and deeper roots. Three to six times the quantity of roots can be produced by grass in several good crops each year as against grass managed, or mismanaged, on the basis of the "constant nibble ', which is the too general present pattern of pasture production.

Further it seems certain that the greatly increased grass production itself will also be of better nutritive quality as well as being more suitable for stock to gather it in. If, as I believe, cattle will graze for a certain fixed number of hours each day and no longer, even when inadequately fed by not being able to collect and eat enough when the grass is too short, the benefit from that system of management for the cattle will rival the benefit for the soil itself. A cycle of improvement for soil for grass and for stock could be the result. Thus good subdivision of farm and grazing properties is seen to be one of the principal factors in the improvement of soil fertility and a major factor in increased production.

It is claimed first that no bases of planning, other than those belonging to the Keyline approach, can possibly satisfy all the basic requirements of rural land development, which must include that of making the best possible use of every natural and renewable resource belonging to the particular property itself. If that be so then it is true that it is a rarity to find any farm or grazing enterprise which has been correctly subdivided; and that without any criticism of the actual fences and their more often inadequate gateways. Therefore the first improvements to be made to the present unplanned or wrongly planned subdivision of land is the general acceptance of the water principles which have already been demonstrated herein. Again it is stated it must be recognised that farmers and graziers can *make their own fertile soil* and thus make good land better, and progressively but rapidly convert poor soil and land to high fertility and more economic production. They will do it as an incidental to the process of production by following the best methods of land and water planning and management, and correct land subdivision by good fences plays a vital role.

On any property the particular past and present climate was and is one of the greatest of all influences on the quality and productiveness of the soil. For instance, if two differently located soils were alike in every other possible respect save only that the rainfall differed, then the soil receiving say 635mm of annual rainfall could be worth from five to 25 times the value of the other land if it had a rainfall of only 254mm. The influence of no other aspect of the climate on soil can be altered and so beneficially changed as can that of water, and likewise no other factor of agricultural influence can be so beneficially altered by proper planning as can water. Water and rainfall are thus the bases of land development planning. No land subdivision has even a chance of being correct unless it has been planned to fit in with and facilitate the planned control and development of all the available water resources.

Further, the planning of the development of land should be based on permanent factors, namely, the twin agricultural permanencies of land shape and climate. The one factor of climate which permits of planning and control being water, so again land planning resolves itself first and foremost into an understanding of the two factors of land shape and the related water movements.

There are various types of natural waterlines of land such as stream courses and water divides. Water flowing over land has pattern lines of flow and predictable path lines of movement, as has been seen. The contour is a waterline principle, and nature's waterline of a like surface is but a true contour line marked out on the land.

And now for the farmer who, when he has understood and accepted the principles of planning and design indicated by those natural waterlines, soon realises that there are other and more artificial waterlines. For example, lines to divert water to fill dams, ponds and lakes of his own making, and cultivation lines to break the natural pattern of water flowing over land to spread it and to improve the usefulness of rainfall which produces runoff. For irrigation, path lines of water movement may be reinforced with earth banks following the course of those lines, as we have seen, to maintain the forward and downland movement of irrigation water on flat land. A true contour water line may be permanently etched on the land surface as a wide irrigation channel which transports water for irrigation and at a far greater volume of flow than that possible on a farm supplied with public irrigation water. A lock-pipe system beneath the wall of a storage dam is likewise a waterline association. It releases water from a storage dam into another waterline — a channel for irrigation. The irrigation channel waterline of flat country differs from the irrigation channel waterline of undulating country. The flat land irrigation channel is built up so that water flows on the surface of the land and the waterline is near to or on the true contour, whereas the irrigation channel of undulating land is excavated so the water flows on a general level which is a little below the land surface, and the water line flow of that channel has a more positive fall, as has been stated, of as steep as 1 in 300.

Again there is another series of somewhat different waterlines which will work in climates which provide rainfall in generally excessive quantities, and where the principal reasons for the interference with the natural waterlines is to improve soil by improving on the natural drainage. Whereas to improve the use of available water resources, the artificial lines for water flow transport the water in the same direction, but on lesser fall, as the natural drainage lines of the stream courses. But in conditions of general excess the manmade artificial lines for drainage may turn the water flow temporarily in the opposite direction.

Now all those natural and artificial waterlines of land are important and of course critically so if the aim of land planning, development and management is the only logical one to make the best possible use of all the natural and renewing resources which are available, including first and foremost those of the water resources and of every kind. But none of the natural and artificial waterlines is a straight line, and yet subdivision fences must be fitted into the waterline scheme of things to even approach the ideal of completely efficient water resources development, together with its effective use and management.

Though fairly wide publicity has been given to the above ideals, the customary methods, or lack of methods, of orthodox land development are still being persisted with despite their very much lower efficiency. It is rarely that one sees even a farm dam for irrigation which has been correctly located, and from which the water is taken out and used by the most rational methods of control and irrigation. It seems that such things are almost never right, not even by accident, despite the fact that they generally follow, albeit haphazardly, and only recently, methods originated on my property more than 25 years ago.

But the fence above the irrigation channel just now discussed is one which does fit in with the best use of water. It is placed in such a way that any servicing of the channel and its banks is not impeded by the fence being too close to the bank. The fence gateways have been placed with proper regard to the water conditions of the valley and ridge shapes of land as well as to provide adequate and workable access.

On any property where its own water resources can be developed economically and used to produce irrigated land, the fence which divides the irrigated land from rainpasture or raincrop land is of the first order in subdivision fences. On stock enterprises it needs to be a good fence, as the condition of the irrigated land is a strong incentive for stock in a rainpasture paddock above it to test its strength. Gates need to be more adequately stockproof than the fence itself, since stock, expecting to be moved, tend to concentrate at the gate and as always those behind push the ones in front. And in the matter of gates the opinion has been formed after much experience that by far the dearest gate is the cheaply made farm gate. The one referred to goes under various names, such as Cocky's gate, bush gate, Taranaki gate, Wagga gate, California gate, and so on. By its continual tendency to fail under stock stress, by being awkward to close and therefore being sometimes left open and by being difficult at most times to open, it causes such a continual loss of time and effort that, by comparison, the most expensively made steel gate would finish up by being far less costly.

The fence above the irrigation channel joins a fence running down the land to form a side boundary of the irrigated land. For instance, if the irrigation water is supplied from an earth wall dam built across a stream, then the stream below the wall of the dam will be either fenced in with the irrigation land or fenced out of it. In either case the fence is located from the waterline of the stream in the same manner that the first fence was located from the waterline of the irrigation channel, and its position is plotted and pegged out in a similar manner. It likewise needs a minimum clear distance from the banks of the stream, and a maxi-

mum distance can be applied as before and gateways located on the same basis.
Again the fence along the stream may be the place for another tree belt which
then, if the creek is fenced in with the irrigation land, locates the permanent
fence farther from the stream as described before in reference to the fencing
above the irrigation channel.

The portion of the irrigation paddock away from the water supply side may
already join up with a suitable fence or a part of the boundary fence of the
property. Another fence must close the lower limits of the irrigation area. Again
that fence could be associated with a stream course. The stream may carry water
suitable for stock and therefore be fenced in with the irrigation area. The fence
then on the lower boundary would be located in the same relation to the watercourse as the other two fences were to theirs. One or both side fences now
should cross over the stream, over a natural waterline, and those fence crossings
like gateways, need to be even more stockproof than the rest of the fence and at
the same time be constructed with regard to the high flows of the stream.
Floodgates may be needed and local design may be the style to follow.

Alternatively the lower border of the irrigation area may have no such a
feature, for the fenceline and the land below the irrigation area simply continues
its gentle downward fall. The fence would then be located by the measurement
of an appropriate distance downward from the irrigation channel, and according
to such factors as the volume of the flow of the irrigation stream, the relationship between the amount of available water and the area of land below it which
lies in the best position for irrigation. Again, when from the available land there
is a choice, it could be decided that the general shape of the irrigation area
should be long and narrow in preference to a squared shape, so the subdivision
within the irrigated land itself could be more simply done on the formalised lines
of selected steering banks. Whichever way the selection is made, the local circumstances and the foregoing principles will make the most suitable position
more or less obvious. The fenceline itself should be located with relation to the
contour, and its final lines be those selected to produce a formalised contour.
Then the fence itself, as well as filling in with and assisting the overall water use
scheme, will serve as a guide for that part of the cultivation of the irrigation area
which parallels upward from a lower contour line.

In practice no disadvantages have been associated with the up and down
fences which produce the subdivision within the irrigated land being constructed
on the steering banks themselves. (See photography *Plate 32.*) However, in that
case the original lines produced by instrument levelling are formalised for the
bank construction, and so the fence is again a series of straights.

The smaller the paddocks the higher is the cost a hectare of fencing.
Paddocks of square shape and with areas of 640, 160 and 40 acres will, with
fencing costs at $625/km ($1000 a mile), cost respectively $15, $30, $60/ha
($6, $12 and $25 an acre) to fence. The differential is actually greater as the
paddock becomes smaller, since the higher cost of gateways is again greater in
price a hectare.

Again as irrigation flows become greater and the irrigation bays between the
steering banks are longer, the subdivision layout of the irrigated land may
require that every steering bank be a paddock boundary and have its own fence.
There is nothing against that degree of subdivision if it pays, indeed there is a lot
to be said in its favor since it simply involves the use of formalised lines for every
steering bank. Further, some of those one-bay paddocks could contain stock
yards and special buildings such as shearing sheds. And what would be wrong

with having a selected irrigation bay as the site for the gardens and orchard of the homestead? Indeed why not place a new house, or even a homestead and all its subsidiary buildings and yards, in such an irrigation bay? There are vast areas of flat land where a hill or a higher site for a homestead does not exist and where a site adjacent to a river, a creek or a billabong is often the only present choice. In such cases a site inside the top end of a specially selected irrigation bay could be ideal.

Most owners would want the greatly improved living conditions which irrigation can bring to the hot areas of generally low rainfall. And many sheep and cattle stations in such areas have major unused and often unsuspected water resources. Of course, when it rains those special irrigation bays would get wet, but now they could be also irrigated at will. Likewise any such area which may be required to be kept dry could be bypassed in the irrigation cycle simply by not opening the watergates which lead into it. Again in conditions of heavy rainfall runoff the watergates to all the other paddocks remain open. Thus runoff from above the irrigation area could be prevented from entering where it may not be wanted, simply by closing two watergates.

Such matters and their management can be easily planned and cheaply executed, provided these factors of water lines and water flows and the general associations of water and the land are understood. The advantages can be outstanding.

While any main subdivision fence which divides irrigated from rainland should be considered in regard to its suitability as a site for a permanent tree belt, there will still be a need for tree belts within the larger irrigation areas which will be developed on many flatter land properties.

Young trees planted following proper soil preparation grow into effective wind breaks much more rapidly than is generally realised. When it is stated that in five years the belt of trees will be an effective something-or-other the impression is often given that from planting time until five years hence a tree is nothing. But a belt of trees at six months and after only its first growth is something of beauty and of satisfaction, and particularly so to the planter and owner. At 12 months the belt of trees already takes on a planned, orderly and well-cared-for look which it liberally bestows on all its surroundings. In a very short time the belt of trees can stand alone and no longer need its protective fence.

But in how much shorter time are those stages of growth reached when, as is now suggested, a belt of specially selected types of trees is planted down the centre of an irrigated area and in its own bay, with its own steering banks and fence boundaries, and where the belt of trees can be watered at will for practically no cost? And there is always a forestry officer in the vicinity with his sure knowledge to discuss the types of trees, their habits, their advantages and disadvantages, and so aid the landowner with his decisions. As well, behind the officer there is his department which will even grow and supply the young trees.

There is the added factor on a property, whether it has irrigated land or not, which has been planned on the proper basis of its own land shape with its associated waterlines of all kinds, that trees can be with surety planted in the correct locations for now, for five, for 20 and for as many more years hence as the farm or grazing property continues as such. Indeed it will be found in some situations that the planting of adequate trees may eventually be the sole reason for the property's continuance.

How and where are trees to be planted for permanent good if those factors of planning are unknown or ignored?

The increasing need for a great deal more subdivision fencing and the rising costs of posts can be offset to a considerable extent by the landsman growing his own fence posts. He can very simply create his own perpetual forest (tree belts) for the purpose.

With the success of the preservation treatment of fence posts by both the pressure and the sap-replacement methods, posts of small diameter cut from young trees are made to last as long as split posts from selected old trees. The smaller treated round posts are usually pointed and driven into the ground. Seedling trees planted for the purpose on my own property provided usable posts in only three years.

The idea is to plant four or five rows of trees, located as discussed for permanent tree belts. The distance apart of the trees in the rows can be as close as 1.5m and up to 2.4m with a spacing between the rows adequate for cultivating with a tractor and chisel plough — say 3-3.6m. Trees can then be cut for posts after three years, when each second or third tree is taken as required. The trees are cut off with a slanting cut about 126mm above the ground. The remaining stump will soon "sucker", and when the new growth is sufficiently advanced one selected stem is left to grow again into a fence post size, and all other stems are cut off. The new tree, with the full root system of the cut tree remaining, will grow very rapidly. By judicious selection of trees for removal and thoughtful management, a valuable shelter belt and a perpetual supply of fence posts can both be attained.

From the national viewpoint Keyline water control systems on both hill land and flat land could make reafforestation on a grand scale both practicable and profitable. It is suggested that some aspects of the Keyline plan offer a most satisfying solution to the evergrowing and problem of Australia's inadequate and dwindling forestry resources.

As is abundantly evident crop land, pasture land and afforestation can be in competition with one another and to each other's detriment. Keyline planning relegates each to its proper locality and function and thereby assists in maintaining high quality production in each case.

CHAPTER THIRTEEN

Cultivation of irrigated land

THE FIRST CONSIDERATION OF THE USE of land in a newly developed farm irrigation project may be other than that which is the best for the land and for the soil itself. For instance, at the time when irrigation water first becomes available for the new area, the main consideration may be the quick wetting of as large an area of land as possible. The matter of ensuring a perfectly uniform spread of the water or the planting of better species of grasses and legumes, both of which may occasion some delay in, say, the production of stock feed, could be of less immediate importance than producing a quick growth even of relatively poor pasture. Diverted water may fill a new dam before rain falls. On the other hand, the dry period may delay any use of the irrigated area until heavy rain falls and runoff fills the supply storage. In that case full use may already have been made of any fodder on the area and the cultivation of the land for the improvement of the soil and the uniform spreading of later irrigation waters can proceed immediately without any disadvantages in lost stock feed.

Cultivation of irrigated land

There are three main purposes of cultivation on irrigated land for grass production. The type, form and timing of cultivation should be such that the three purposes of cultivation can, where appropriate, be accomplished at the one time and at each time when cultivation is required.

The three purposes of cultivation on any Keyline project are (1) as an aid to the development and continuous improvement of the soil, (2) for the uniform spreading over the soil of water both from rainfall runoff and from irrigation — that purpose being contemporaneous with (1), and (3) for the preparation of the land for seed sowing or the actual cultivation when the seed is being planted.

In (1) the mechanics of the cultivation should be such that the main body of the active soil depth is left, as a result of cultivation, in a much improved condition of aeration and that the limited capacity of soils, most of which are dense or heavy or have become compacted, to "take" water quickly is also greatly improved.

In (1) and (2) reference is made to Keyline cultivation and the geometry of Keyline already discussed elsewhere.

The main purpose of (3) is often a purely mechanical solution of the problem of controlling the present growth by killing it to successfully plant and at the same time promote the growth of new plants. But a cultivation which is successful from that point of view can, as a continued process, progressively deteriorate the soil. However, in the case where the planting of new pasture species into a sward which is somewhat less than satisfactory is the appropriate course, the three purposes of cultivation may be fulfilled at the same time by planting the new seed with a chisel-seeder and by adjusting the depth of the cultivation to suit the present stage of the development of the soil, and by placing the seed at a depth in the chisel furrows which best promotes its germination and growth.

Cultivating equipment

Each of all the types of cultivating implement has its particular uses and special advantages and disadvantages. The mouldboard plough is considered the best implement with which to mechanically prepare a "clean" seed-bed when the soil is moist and the prime consideration is the quick, if temporary, control of the grass and weeds of the cultivated land. By precisely turning the soil and the growth upside down, it thus presents a "clean" soil surface into which crop seeds can be sown easily. It would be hard to find other favorable aspects of that piece of equipment for irrigation land.

The disc plough in its lighter weight versions, such as those combining seeding equipment, has its main advantages in somewhat drier conditions than those mentioned for the mouldboard plough. It promotes a very fine cultivation and the precise planting of grain on the "firm" ground below or at the "bottom" of the cultivated soil and at the full depth of the very shallow cultivation. With the seed on a firm bottom or bed and covered with a fine to a dusted soil which then forms an effective barrier against evaporation, the available moisture is retained at the "firm" seed-bed and so germination and first growth is promoted. However, the "dust-mulch" which retards evaporation also forms equally as effective a barrier against the favorable entrance of rain when it falls later. The finely cultivated soil then may quickly seal in any shower and the critical air movements within the soil are immediately and seriously affected. In those conditions the disc plough can be a fertility destroyer.

The heavier and bigger forms of the disc plough are very effective from a purely mechanical standpoint in their ability to "chop up and turn under" various types of light scrub growth, and to do the work with speed and reasonable economy. The later natural processes of the disintegration of the plant material turned under generally takes a long time and when it does finally decompose there is usually little benefit to the soil. Although almost any other method of bringing such poor scrubland into production would be better for the eventual development of real fertility in this soil, the implement, by being mechanically successful, does have prospects for continued use in such conditions.

Continual use on the same soil of either of those two implements, the mould-board or the disc plough, will in general eventually reduce the fertility of the soil. Their use in areas of good soil and adequate rainfall, in climates much cooler than tropical, can continue for a long time without any notably bad effect and, of course, really good farmers in those conditions can hold and some may even improve the fertility of their soil. However it may take exceptionally good soil management practices and the use of special grass and clover mixtures to prevent rapid soil deterioration when they are used in humid and tropical conditions. Cultivation procedures invariably become more and more critical for soil fertility as the rainfall and the temperature are higher.

Rotary hoes again have their good and bad features, In conditions of heavy weed growth on intensively cultivated soils, such as those used to grow garden and truck crops, the rotary hoe will mulch the growth and incorporate it into the top soil in a very efficient manner mechanically. However, if used on land with poor clayey soil and with little growth they produce a fine cultivation which seals immediately when rain falls thus drastically affecting soil aeration and soil fertility.

The tyne-type implements, ranging from light cultivator to the strong and "go-anywhere" chisel ploughs, are generally much less destructive of soils no matter how inexpertly they are used. But when used in a manner which takes proper heed of soil development and fertility, they become an outstanding and economical tool for soil improvement. However, in the moist conditions mentioned as the outstanding ones for a mouldboard plough, the use of tyned implements for sowing wheat would be unlikely to result in as heavy an immediate crop, but neither would their continued use be nearly as damaging to soil fertility. They are more likely to improve the soil and within a short time produce better food-quality grain at a more economical cost.

The outstanding way to convert a poor earth into a highly fertile soil, and in a short time, is through the proper development of pasture, and here the chisel plough is the outstanding implement. And in the development and continual improvement of the soil of irrigated land the chisel plough is supreme. Even if another cultivating implement is at times used, (the damaging sidecasting implement such as the mouldboard and disc ploughs are better excluded from use within the irrigation land) the special cultivation should be followed immediately by a cultivation with a chisel plough.

The depth of any cultivation, including the use of other implements as well as the chisel plough, must always be determined in relation to the present particular conditions and depth of the soil.

Keyline cultivation

The cultivation pattern for the uniform distirbution of water is dealt with elsewhere. Nevertheless it is appropriate to restate some of the techniques of cultivation here in the light of the discussion on flatter land irrigation.

The shape of the irrigation channel gives an immediate clue to the shape of the land of the irrigation area. If the channel is close in form to a straight line the land shape is extremely large and very uniform. In such conditions a contour line halfway down between the irrigation channel and the lower end of the bays and one other along the lower boundary of the irrigation area will in all probability follow the line form of the channel.

A chisel cultivation of that land done either parallel downward from the irrigation channel or from any other contour line, or parallel upward to the irrigation channel from a lower contour line, will both have the water-spreading effect desired. And such cultivation done on pasture land immediately after it has been eaten off leaves most of the old pasture intact and the soil in the best condition for immediate irrigation.

However, where the irrigation form has produced curves which swing both upward towards the rise and downward towards the fall of the land, the curves clearly indicate the presence of both valley and ridge forms, even though the forms can be so flat as to escape the notice of the naked eye. But the water flow will be affected by the shapes, and if no measures are taken to spread the water it will follow its natural path somewhat sideways from the flat ridge form to the flat valley form, to the extent that it can do so within the limits of the steering bank sides of the irrigation bays. While a good spread of water may still result without Keyline cultivation on land of little fall which is also uniform, cultivation will nevertheless improve that factor. But it is still necessary for rapid soil and pasture development. Then, the pattern ploughing of Keyline cultivation will simultaneously consolidate the continued even spreading of the irrigation water, improve both the soil and pasture and, if desired, new seed may be sown into the present irrigated sward with the cultivation.

In the circumstances which are clearly shown by the above cited curving shapes in the length of the irrigation channel there are only two patterns of cultivation to employ: (1) The valley-shaped areas which lie within and below the upland swinging curves of the contours are cultivated parallel downward, and (2) the ridge-shaped areas which lie between the downward swinging loops of the channel and the lower marked-in contour lines are cultivated parallel upward from a specially marked contour line or lines below. The effect of those cultivation lines and furrows is to somewhat oppose the natural flow path of the water and so cause a uniform spread of the flowing water.

For cultivation, a contour line is pegged in and marked with a suitable furrow across the area in the position stated. A similar line is placed near the lower boundary of the irrigation land, or a contour fence line serves the same purpose. Those guideline furrows should be prominently made with a channel-maker, a delver, a ditcher or a large single-furrow plough, whichever is available. It is desirable that they should be easy to follow for several years of the use and management of the irrigated land. They should not however be of such form as would be an obstruction to station implements and vehicles.

The preservation and usefulness of the lines can be enhanced by sowing a special grass into the newly made furrows. Another simple means to the end of preserving the marked contour lines is that of leaving a 1.2m to 1.8m wide strip of land in a permanently uncultivated condition on one or both sides of the furrow, or where a large area is being dealt with, to leave a strip of land out of cultivation and wide enough for a farm road.

If it suits the stock position at the time water is first available, the whole of the prepared area should be Keyline cultivated to the appropriate depth according to the soil conditions, and then immediately irrigated. Preferably the man who does, or is in charge of, the cultivation should be the one who does, or is responsible for, the irrigation. He should constantly follow the water as each bay is irrigated the first time and note carefully the spread of the water. Preferably he should carry, as well as a shovel, a few pegs, and place them to preserve the location for later correction of areas of less effective spread. The performance of steering banks and the effectiveness of the stops in the small excavated drain left in constructing the steering bank are kept under check and adjusted as required. It is important that all the construction work be fully effective. Water should always be kept under complete control. No leak even of small proportions should be permitted from the irrigation channel. Watergates must not only release the irrigation stream effectively into the irrigation bay, they must also shut off every drop of water when they are closed. No leak at a closed watergate should be permitted.

The steering banks likewise must do more than just their part in steering the water down an irrigation bay. They must also ensure no water returns to a bay on which irrigation is just completed, and that no water escapes ahead of its time into an adjacent bay on which irrigation has not yet started.

The landholder should also check the time required to cover each bay and record it against the number of the bay for the first few irrigations.

When for the first few irrigations, the patrol and reconstruction work is done, and the times noted, later irrigations could be completed without moving from the immediate vicinity of the irrigation channel and its watergates. Thus, the only work in irrigation is in walking along the irrigation channel and opening and closing watergates according to the time record.

The amount of irrigation water which one man may effectively control in flood-flow irrigation is therefore finally limited only by the number of watergates he can open and close in, say, half an hour. If he can open and close 10 watergates in that time over a distance along the irrigation channel of 230m, what is there to prevent him irrigating 15ha each hour, provided always that enough water has been collected and that the area of land is large enough for that rate of irrigation?

Obviously there will be limiting factors which will govern the final irrigation rates, but the limits are much more likely to be imposed by factors of water and land — and long before even a single man's capacity to control water and irrigate land has been reached.

In the cultivation and management of irrigation land it should be constantly kept in mind that there are hazards to be avoided as well as great advantages to be gained from irrigation. The good, complete and fertile soils of nature are not associated with the higher rainfall areas but with those where the rainfall is from 406-762mm a year. The addition of the equivalent of an extra 457mm of rainfall as irrigation water is thus more than sufficient to cause serious soil leaching, and particularly so where the climate already provides 508mm or more. The overleaching of essential minerals from irrigated land can only be avoided by the strict control and limitation of both the amount of water which goes into the soil and the time during which the water is in contact with the soil. It is essential therefore that irrigation procedures should water the land as quickly as possible.

The speed of water application in both Keyline pattern hillside irrigation and flood-flow flat land irrigation is such that both hazards to soil fertility and soil improvement are completely offset. Indeed it could appear that the completely opposite effect of too little water entering the soil at each irrigation could be a problem where soils are generally compacted and infertile. However there is always very rapid water absorption for the first 51mm or so of water into highly fertile soil, and into compacted and poor soils when they are appropriately chisel cultivated.

The objective of irrigation and the result or irrigation should always be the improvement of the soil.

CHAPTER FOURTEEN

Fertility —
a product of management

OVER A QUARTER OF A CENTURY ago the father of American soil conservation the late Dr H.H. Bennett wrote in his book Soil Conservation:

"Lack of foresight and restraint . . . has created in this country a land problem of tremendous implications. What makes the situation so grave is the irreplaceable nature of soil. Once this valuable asset leaves a field, it is as irretrievably lost as if consumed by fire. Soil is reproduced from its parent material so slowly that we may as well accept as a fact that once the surface layer is washed off, land so affected is, from the practical standpoint, generally in a condition of permanent impoverishment. As nearly as can be ascertained, it takes nature, under the most favorable conditions, including a good cover of trees, grass, or other protective vegetation, anywhere from 300 to 1000 years or more to build a single inch of topsoil! When seven inches of topsoil is allowed to wash away therefore, at least 2000 to 7000 years of nature's work goes to waste."

As recently as April 1965, American Agricultural Trends, which is compiled and issued by the United States Information Service, contained 16 items of agricultural interest. The last, under the heading Small Sheet of Topsoil Sustains World's Life, read:

"All the world's human life depends on the fertility of a thin sheet of the earth's topsoil, covering 1/10th of the earth's total surface and forming a storehouse of plant nutrients only about seven inches (17.8 centimetres) deep. It takes about 1000 years to build up one inch (2.5 centimetres) of topsoil. Wind or water can take that much away in a few months unless the surface is protected."

FARM DAM CONSTRUCTION

PLATE 34 — *Site preparation with the cutoff trench completed. The lock-pipe trench crosses the cutoff trench at right-angles on this side of the tractor.*

PLATE 35

Laying the lock-pipe system. Note the baffle plate halves set in the ground between the men. (The front man is positioning the top half of the baffle plate.) The bulldozer is in the pond area of the dam.

PLATE 36

Wall under construction. Note the drum which stands at the open end of the lock-pipe.

PLATE 37

Finishing off a farm dam.

203

ROTHSTEAD

PLATE 38

Sir William Ogg, left and Sir C. Stanton Hicks with the 1899 Rothamsted U — wheat grown on the Braodbalk strips.

PLATE 39

Illustrates the first and the fourth year after fallow for wheat grown in 1943, — See text Chapter 17.

PLATE 40

Stream diversion. Portion of an excavated channel which diverts a small stream to supplement rainfall runoff to keep a dam on a grazing property filled for flood-flow irrigation.

PLATE 41

A dam on a ridge. High Circle dam on Nevallan is pump-filled from a creek dam below. Equipped with a 330 mm (13 inch) lock-pipe system it provides water for both the stockyard and for hillside irrigation.

Planted tree belts. A scene on Nevallan with part of Yobarnie in the left background. The planted tree belts on Nevallan are principally tallow wood and spotted gum supplied by the New South Wales Forestry Department. The picture is by courtesy of the department which used a similar scene of Nevallan for two of its tree catalogues.

PLATE 43

Levels are critical. The Bunyip level, designed by the author, in use to mark a line for an irrigation channel. The transparent tube is 60 feet long and five feet of each end is held in the extruded aluminium staffs. When filled to within a foot of each end with water the relative heights of the staffs can be read-off on the water level on each staff. The 50 feet of "slack" tube between the staffs is used as the only measure required for such work.

PLATE 44

A closeup of a watergate installed in the bank of a small diversion channel pictured during an advanced Keyline school on Yobarnie. Water from rainfall runoff flowing in the channel can be released to fill a lower dam or to the creek when the higher dam it serves is filled.

Those two statements made over a quarter of a century apart, represent only too well the attitude to soil of agriculture's officialdom. Is it any wonder that, when such hoplessness is official, and mistaken conceptions of soil still so widely promulgated, that farmers and graziers are adversely affected and their land suffers needlessly?

Wrong thinking on soil constitutes the greatest of all obstacles to the proper development of the farm and grazing landscape. I believe it to be a truism that the development of land depends in the first instance on how the farmer thinks about his soil; and if he thinks rightly about his soil he has a chance of doing the right thing by his land. But if he doesn't think correctly about his soil he won't have a chance of doing the right thing by his land — not even by accident.

The purpose of this chapter is to infuse some realistic optimism into the manner of thinking about soil and at the same time to establish a practical and enlightened foundation on which to support the various themes of this book.

The mineral world of the soil is composed of the dust and rock debris of the decomposing rock surface of the earth. The process of rock disintegration, the movement of the detritus from place to place by all the natural forces, the reassembly of the particles into later rocks of different character which were again decomposed, the addition to the original materials of reorganised mineral products from various forms of past plant, animal and sea life and the constant classifications of all the various dusts by the agencies of wind and water, has been going on for many millions of years. Moreover, much of the present agricultural land of the world has been once, and some of it several times, beneath the oceans.

That is the first part of the natural soil forming process and it is a very slow process indeed. But the time has passed, millions on millions of years of it, and that material — the basic soil-material — now exists in relative abundance under most of the agricultural lands of the world. And that is almost universally so whether the topsoil has been lost through man-caused soil erosion or not.

The second part of the soil forming process — the biotic or organic — by comparison, is certainly extremely rapid in any climatic conditions which permit agricultural pursuits. The time requirements in the second process relate more to the periods of the life cycles of the various soil communities, of the grasses and plants, and of the animals and birds which invade the soil in search of their food.

In the second or rapid soil forming process man can interfere and so control and speed up the various biotic and soil climatic forces involved that he can transform the plentiful inert soil-materials into the complete world of the soil in as little a time as two or three years. And beyond that period the soil will continue to improve according to the manner of the thought and the care it is given, and as part of the general management of each farm or grazing property.

PRACTICAL EXPERIENCE WITH KEYLINE

An early experience of mine in this type of soil making had as soil-material, yellow clay subsoil (from which the top 76mm of poor grey soil had been washed away), soft yellow shale rock and a harder blue shale. The original profile had been 51-76mm of grey soil, about 229mm of yellow subsoil on the hillside ranging to a metre or so of the same material in the valley areas, all underlain by the harder blue shale which was the immediate geological base of the land. The areas of yellow subsoil clay were associated with the lower parts, the yellow shales with the medium slopes and the blue shale showed on the steeper ridges.

We cultivated the land 51-76mm deep with a chisel plough (which does not turn the soil upside-down) equipped with 51mm chisels spaced 305mm apart. The first cultivation was done in such a way that the second run with the chisel plough, when made on the Keyline pattern (see Chapter 6) would form a cross cultivation. The second run was made slightly deeper than the first. Into the rough and lumpy surface was planted a mixture of grass and inoculated legume seed together with one kg a hectare of a 50-50 lime-superphosphate mixture. With the planting, white clover and cocksfoot were included and both were said not to grow in the region. The weather was warm and the moisture conditions were not critical. The result of that sowing was exceedingly poor. Very few clover plants remained for long after germination, and the grasses were thin and yellow.

That "pasture" was allowed to grow to about 178mm high when cattle were put on and they ate it right off. The area was then locked-up and the grass allowed to grow as before and again eaten off. The paddocks were finally eaten out in the early autumn by which time the grasses had improved somewhat in both color and growth, with a thin sprinkle of white clover and lucerne plants persisting and many more and better plants of subterranean clover. It was immediately chisel-cultivated on the Keyline pattern a little deeper than before. The result of that cultivation appeared to be almost complete destruction of the indifferent pasture. However a notable change was evident in the soil-material after only 12 months, with the lumps more friable and a darkening of the yellow material. The most notable change had perhaps occurred in the harder blue shale where at first the plants had appeared to be scarcely alive and certainly were not expected to persist through to the autumn. There was now much more finer and darker material while the broken shale and the larger shale pieces were a mass of fine cracks with many of the lumps fretting away and breaking up when touched. The persistence of the various plants was then about equal to that on the yellow areas.

Pasture needs its time to grow

The paddocks were left without stock for the pasture to grow. Regrowth was relatively rapid, the color was much improved and the change in the clover was notable. While the subclover had thickened up, the white clover was now everywhere evident. Grasses grew taller and thicker. No more fertiliser was used then or at any time during the next 10 years.

The pasture was given its time-to-grow after each stocking during the second year when it was cultivated in the autumn as before, but again a little deeper.

The "soil" had by this time been transformed. For a start there was little evidence to be seen anywhere in the rough cultivation of the former yellows; the material was now much darker. The blue shale patches looked remarkably like good dark soil. Many small earthworms were to be seen in all the cultivated and aerated "soil", and throughout the second year the pasture itself had continued to improve but at an accelerated rate on the improvement of the first year. Root systems were also deeper and very much thicker everywhere. Many visitors who saw the pasture at the end of the second year believed we would ruin what they then considered was a good pasture if we cultivated it so drastically again.

But we had seen the rather spectacular improvement in growth after the apparently destructive cultivation before, and so after the third year the paddocks were again chiselled a little deeper. Throughout the third year the pastures were very good. The rate of recovery from second year chiselling had further improved.

The fourth year pasture was outstanding. The soil everywhere was then dark and teeming with soil life. While there were many more small earthworms there were also considerable numbers of larger ones.

From the start of that particular soil development we had dug samples of soil with a spade every Sunday afternoon and continued the examinations each Sunday for many years.

The cultivation program was completed at the end of the third year when there was an average of 127mm of real soil over all the land which had been treated. But the soil continued to deepen and become darker as the earthworms increased in numbers and in size.

Soil transformation

The new topsoil transformed from the soil-material was, we were convinced, deeper, darker and far more fertile than any which had ever existed hereabout in existing climatic conditions.

That program or similar ones have since been repeated on the widest variety of soil-material; on poor soil and on soils not so poor, all over the farming and grazing lands of Australia — and with similar results.

On many occasions property owners have started the program and obtained such outstanding results in improved pasture — better than they have ever achieved before — that they were afraid to continue for fear of losing what they had gained from one one chiselling of their old pasture and soil. But the soil cannot be permanently improved by only one year's care.

In his book The Geographical Basis of Keyline, J. Macdonald Holmes, Emeritus Professor of Geography at the University of Sydney, observes of just such a soil program: "The soil he dug for us was a revelation . . . the three year course of Keyline cultivation had developed an extraordinary depth of dark soil just teeming with soil life, while the pastures were thriving."

In review what we had done was to place the initial soil-material in a condition which allowed it to take in rapidly more of the rain which fell on it and so to remain in a moist condition with near perfect aeration for long periods. By using artificial fertiliser as a trigger element and the lime to provide the alkaline medium so necessary for clovers at germination, we had caused plants to grow. We had allowed the grass growing time to stimulate root-growth. Much of the root organic material died soon after the grass was eaten, so forming food for life in the soil. So with good living conditions — moist, warm and well-aerated soil — and a plentiful food supply (the recently dead roots of grasses and clovers), the soil species available from the atmosphere and the surroundings and from the rumen, glands and other organs of the stock, developed rapidly to a climax.

The change which only one such climax of soil life will bring about in the look and feel of a soil is quite dramatic — for those who would see — but generally no one is looking. Too soon, however, the effect on the soil of that one climax will be lost. The soil will drift back slowly to its former condition. But

in that instance we caused climax to follow climax each time the grass grew up
as the roots went down to the limit of the depth of the aerated medium. Each
time the roots died the aerobes of the soil responded to the good living conditions and to the food supply, by breeding frantically. Each climax of soil life
development ended with the death of countless millions of the soil inhabitants
to play their further part in death, in the enrichment of the soil.

The Keyline cultivation done in the early autumn of the first, second and
third years was made a little deeper each time. And each time the cultivation
again increased the moisture holding capacity and maintained the best possible
condition of aeration. So each time the grass roots grew deeper and to the full
depth of each new cultivation, and increased in quantity as the soil improved
rapidly.

Succession of climaxes created

As has been indicated the soil change brought about by one of those climaxes
may not persist for very long but here we had promoted a rapid succession of
climaxes.

There were up to five climaxes each year, following each eating out of the
well-grown grasses. And there were three greater climaxes which followed the
autumn chiselling. The result was, and always will be in like conditions, a transformation of the soil which is relatively permanent.

A volume a dozen times the size of this book and written to explain all the
soil processes involved in the making of fertile topsoil, would inevitably leave the
greater part of the story untold. But we did not know the complete story and
never will — we had discovered only how to create the situation rapidly.

While it is no more necessary to know all about the soil processes to manage
soil and to improve soil than it is to know all about electricity to switch on a
light, we do need to know somewhat more than the obvious items chronicled
in this experience.

Actually we can know very little from this description, since the climate has
not yet been defined. The climate belongs to that of a mid-temperate zone,
64km from the sea, from 30 to 91m above sea level, average annual rainfall of
660mm — no reliable incidence since it lies nearly midway between the southern area of strong winter incidence and the northern areas of summer rainfall.
There is a wide variation in the yearly rainfall from a low of under 178mm to
a high of nearly 1778mm. Summer temperatures occasionally reach 37.8 deg
C with an average maximum around 29.4 deg C. Winters provide only a few
heavy frosts.

CRITICAL SOIL CLIMATE CHANGED

It has been said a particular climate is a dominating influence in the formation
of natural soil, but here it has been shown that the effect on soil of the regional
climate can be manipulated to change the critical soil-climate and so produce a
better soil than the earlier natural one.

Even with the climate stated it can be seen that it is necessary to have a sufficient knowledge of, and a feeling for, the principles involved in this soil-making
so as to be able to apply those principles to places of different climates and soils.

The question is, could those techniques be followed willy-nilly in the same
zone but in areas with an average annual rainfall of 356mm and a winter incidence, and in another with 1270mm and a summer incidence? The principles in-

volved relate to the sowing and feeding out of pasture, and the treatment and management of the soil, all in such a way as to produce the maximum in the vital soil climaxes and sufficient numbers of them one on the other, to permanently (relative) improve both the soil and the pasture, and soon thereafter the whole landscape.

For a start the cultivation of soil in autumn as we had done could be quite unsuitable in both those other cases. Chiselling is to provide good aeration of the topsoil plus a little of the subsoil at the time of the year when, following the cultivation, the moisture and warmth conditions in the soil are most likely to be best suited for climax promotion.

Techniques vary with climate

Soil influenced by the low winter incidence rainfall would lack moisture in the warm months of early autumn, and when the rains arrived later on the low temperatures would inhibit climax development. Moreover, pasture for winter feed would be depleted by the cultivation which would worsen the conditions of overwet and boggy soil. But here a cultivation immediately on the first eating-out of one or two paddocks in spring may promote the working of those dynamic principles.

In the somewhat warmer climate with its 1270mm of predominantly summer rains, a spring or autumn start on a soil improvement program may be pointless. Of course, the best time to complete the first "soil aeration" could be the day before it rains. But who knows when it is going to rain? However a paddock or two may be devoid of grass and so dry there can be no more growth until rain falls. As nothing can be lost all must be gain. Whenever it rains for those paddocks the cultivation will have been right.

But what about the changes in procedures that may be necessary for differing soils, as well as those which may be dictated by the various climates? There are two principal aspects of soil to be considered, one which can be readily seen and the other unseen.

First, it is obvious whether the surface is a soil (no matter how poor) or merely a basic soil-material. If it is soil, then there are some aspects of the cultivation which are affected or governed by the features of the soil. Now considerable areas of agricultural Australia are covered by only 51mm of top soil which is alive, but there are also parts, notably in the poorer granite belts, where the living depth of the soil is little more than a smear of darker color over the very light colored material below. Then again some soils are deep but are still relatively infertile and low producing. Each of those soils requires a somewhat different treatment to that given to our soil-material.

Second, the unseen relates to the necessary mineral elements of fertility and to their availability. Some of the elements may be so scarce that healthy plants — and that always includes the animals which are forced to live there — are impossible. On the other hand all the minerals may be adequate or even plentiful, and be still useless to plants by being unavailable in their existing form. They then need further breakdown by processes not very well understood, but which always are present in and a part of every soil.

SOIL-MAKING PROCESS ACCELERATED

Referring back to the first of the two aspects of soil, any present fertility — be it only a smear or millimetres deep — can be used to great advantage in the planned process of soil management, by regulating the depth and the type of the cultivation accordingly. For instance if a soil 51mm deep is cultivated 102mm deep with a mouldboard plough, the live soil is buried under 51mm of subsoil. None of the soil would be visible. The soil species and communities may be cut off from their adequate air aupply and their general development greatly disturbed. Further, since it is the intention to use the live soil for a kind of "yeasting" conversion, too much dead subsoil may overtax and slow down the rapid process. On the other hand a chisel type cultivation done to 76mm total depth would leave the topsoil where it should be. It also promotes aeration where it is needed, accelerates the vital processes of the soil and leaves the smaller section of disturbed subsoil in the ideal position and condition for roots to extend into it; and so the earth is brought into the whole process of conversion to topsoil. While other types of cultivating equipment may have their uses, the chisel plough with its "scratch" type cultivation is superior for such soil work.

The position is changed again in the granitic condition recently mentioned. Here it would seem that any depth of cultivating could be too deep. Ideally, chisel spacing could be very much closer and penetration depth at the start somewhat shallower, if the little soil is to be used to the best advantage. But in that type of soil and sandy soils generally, there is less need for deeper penetration to provide air. Any lack of air in those soils is usually caused by a fine surface seal. With that broken the air penetrates readily through the sands which usually are found just below the surface soil.

Depth of cultivation sometimes critical

The depth of the cultivation in similar conditions to those of the early experience quoted are not so critical. Since there is no soil to use, the depth is regulated to those quoted by the fact that too deep a penetration would induce rapid waste of the soluble fertiliser downward, and so reduce the all important initial growth of the new germination. A cultivation deeper than necessary also increases the cost of the work.

When soil-making can be so rapidly and cheaply achieved there is no need to be pessimistic about soil. While it is true that only 1/10th of the earth's surface is covered with soil, and not too much of it could be classed as highly fertile, it has to be realised that agriculture itself can be a soil-making process.

It may appear that the farming and grazing of land has resulted thus far in the destruction of much more soil than it has made. And the many books written on the ravages of soil erosion would seem to confirm that is so. But all soil can be improved, and the relatively larger areas of poor soil can not only be made fertile but the soil process can be greatly deepened, and thus bring into the productive processes of soil more of the nutrient elements than are at present of use.

CHAPTER FIFTEEN

Truths on trace elements

LIKE ALL ASPECTS OF AGRICULTURE one factor affects others. Mineral deficiencies, if any, and including those needed in only trace proportions, are usually known in a general way for all the agricultural soils and through departmental advisors or agricultural department extension services. Those minerals would need to be added as part of the trigger elements at the start of an improvement program, but once the soil-forming process is satisfactorily started they may never be needed again. Many of the poor soils and soil-materials, which are reputed to be short of one or several of the mineral necessities, do contain them. They may at first be unavailable, but because they are recommended and used and nearly always show an improvement in growth or in the color of growth when they are used, the fact of their natural presence in forms unavailable to plants is sometimes disguised.

The visual evidence of the effectiveness of an added mineral or added trace element does not in any sense prove its need or that the soil is lacking in it. That fact is completely ignored and often not even suspected by many extension workers.

Side by side with experience of ours of subsoil and soft rock conversion to fertile soil, another area was treated in an exactly similar manner except in one respect — there was added with the seeding every trace element recommended for any area extending well beyond our soil zone and climatic setting. As would be expected the first growth from that trial with "all added" was better than the other. First and most notably, the color of the pasture species was by comparison good. Second, the rate of growth and the thickness of growth was a little improved and both those effects showed for a little over a year. But about 15 months after the original application there was no difference in the results to be seen between the two programs.

I recounted that experience later to two separate groups of CSIRO scientists during their visits to my property. Naturally they wanted to see for themselves and after it had been pointed out that the "trace element paddock" was one of four visible from the discussion point, an inspection was made. The inspection included the taking of many samples of soils dug with a spade in the various paddocks. At the end of a two-hour walk no one was prepared to make a firm selection of any paddock as the one treated with all elements. Now an agricultural scientist is not necessarily expert on such matters but on the second of these occasions the late Sir Ian Clunies Ross was the leader of the party, which included Dr John Anderson, the trace element discoverer and research authority whom I had found was at least as wellknown and respected overseas for his work in that field as he is in Australia.

Those happenings are recounted to add emphasis to what has been already said on the matter of minerals of all necessary varieties and quantities. The fact that when used they cause the appropriate responses in the treated pasture or crop is not by means proof of their shortage in the soil or in the soil-material.

Availability of primary minerals

The "soil" of that examination which was not treated with trace elements showed by the poor color of the grasses and by other signs in the first year that there were deficiencies in the soil, but less than two years later there were no plants with deficiency signs. The course of the soil-making program, by triggering off vital soil processes, had changed the minerals which were always in the soil and subsoil into forms suitable and now available for plant nutrition. However, that does not indicate it is not wise to add the trace and other minerals to the "trigger" fertiliser at the start of such a program. Indeed it could be safer to add them, as had the conditions of climate been somewhat worse, any means which ensure a better start could play a critical role.

The whole story just illustrates one of the widest held fallacies of both official and commercial agriculture, namely, that if deficiency signs in crops and acid reaction tests of the soil indicate one or more deficiencies, then there is only one corrective course to pursue; it is assumed that the particular minerals must be added artificially to the soil.

The diluted acids used in soil tests do not disclose the great truth that soil minerals can be at one and the same time (1) insoluble in both water and the test acids and (2) available to the plants by other soil processes. Indeed if nutrient minerals could not be both "insoluble and yet available" they must have all been leached from any worthwhile agricultural soil long ago, since only the lack of rainfall would have allowed them to be retained in the surface soil or shallower sub-soils.

When soils are deficient so plants growing in them cannot obtain their proper mineral requirements, there may be two means and not only one by which they can be supplied. Minerals may be (1) added to the top-soil in the same way that superphosphate is used, and (2) they may be made "available" by mechanical treatments of the soils which have always contained them but which are now in a locked-up form. With regard to soil-contained minerals, they may even be absent from some shallow top-soil yet be in adequate but now unavailable supply only 25mm or so lower down in the sub-soil itself.

Because of the tremendous importance of this particular factor of soil improvement it is worthwhile pursuing the matter by relating two other aspects

216

of those visits by the scientists. During one discussion I was asked if I believed in trace element additions to the soil. My reply was in the affirmative, but I suggested that from the present manner of their use it could not be known whether they were unavailable but actually present, or absent altogether from a particular soil. For instance, trace elements were being added over indifferent pastures growing on very shallow and perhaps woefully depleted soils. When the "missing" elements had been determined by trials and then added, the results were quite spectacular. But it could not be known from the recommendations given to farmers and graziers on the use of trace elements, whether the trace elements could be made available at less cost by other and more soil beneficial means. Could it be that the trace added, but always at considerable cost, was still far more than a soil required?

To that point in the discussion, of which only the briefest version is given here, Sir Ian Clunies Ross had been a listener only. He now asked how I would suggest the work be done. My reply was on these lines: Practically all soils relevant to the discussion were poor or depleted. Nearly all Australian aged soils were "soils-in-decline" and long past the stage of their highest fertility. Therefore it should first be determined what each soil could do for itself in the way of improvement, after treatment and management aimed solely at improving the moisture-air relationship in each soil, or, improving the soil-climate.

All soil men highlight the importance of soil aeration. But by comparison with what was done about it and with what we had done and what was now suggested, ours was truly a very drastic soil areation. For instance, with pasture on a 51mm (2 inch) soil, we would literally tear it to pieces with a chisel plough, but digging no more than 75mm (3 ins) deep. The cultivation would be timed to suit warmth and moisture conditions. Then the soil and pasture would be managed as we had done before by giving the pasture its time to grow after each eating off. The same timed cultivation once each year for three years would be given while continuing that same pasture management.

At the end of that time the soil would have available to plants at least a goodly portion of the minerals which it did contain. On the "climate" improved soil all the trace element tests and experiments could be conducted. My opinion was that the tests would then tell the truth.

In support of the program there would be no need to delay any pasture improvement work which now involves the application of trace elements. But in as many of the cases as possible small areas, not "plots", should be kept free of added elements and subjected to the "soil-climate" treatment only. Sir Ian at that juncture remarked to Dr J. Anderson that they should do some of their trace element experiments over again and on those lines.

CSIRO ATTITUDE TO KEYLINE

After that visit Sir Ian continued to display his interest in Keyline generally and to the extent that he later asked would I assist with the Keyline planning and consulting if the CSIRO allocated for Keyline development an area of 240ha (600 acres) on the new land to which the Canberra experimental farm would be moved in a year or so. That was, of course, some years ago. My full co-operation, to which one condition was applied, was accepted then by Sir Ian. The condition was that no soil conservation philosophy or practice should be allowed to intrude on the selected area, so the experiment could also function in the nature of a "test-of-Keyline"; surface and gully erosion which were both present on the land then was to be "cured" as they are in Keyline, only incidental to the Keyline work.

In company with officers of the CSIRO in Canberra I visited the selected area on more than one occasion. However with the death of Sir Ian the Keyline project did not get started.

The other matter of significance in mineral deficiency and mineral supply relates to the first of those visits by CSIRO scientists. In the party were about 10 CSIRO officers, two bankers and a grazier. The usual "farm-walk", with its many examinations of pasture, the digging of numerous soil samples and the discussions with questions and answers on the spot was over, and we were at Nevallan homestead having afternoon refreshments.

One of the CSIRO officers addressed a question to me on the lines of "What would you do if . . .". That the question had some significance to all or indeed that they may have been waiting for it, was apparent by the immediate quiet of all voices save the speaker's. And he continued to relate that a certain area of 1600ha (4000 acres) of very poor soils was "pasture improved" by the now orthodox means, but which were new techniques at the time the improvement started. Clovers and grasses were sown and superphosphate was spread year after year for many years. Initially very few sheep could be carried on the unimproved land but soon it carried 2.5 sheep/hectare (one sheep to the acre). That had never been done on such country hitherto.

The carrying capacity continued to rise as the grazier prospered; and so a carrying capacity of 12 sheep/ha (five sheep to the acre) was reached. That was one of the first properties in the particular State where the reccommendations of the scientists on pasture improvement had been faithfully and enthusiastically followed. The outstanding success here had given pasture improvements a new impetus among graziers of sheep in the surrounding area.

But after a number of year the now consistently high wool clip suddenly and unaccountably dropped by 10 bales in one year. However the grazier, though baffled, was not unduly concerned by one year's reduced clip. But within the next year utter collapse occurred of the ability of the improved pasture to carry sheep. There was a disastrous loss in sheep which were dying in large numbers, necessitating the quick removal off the improved country of all the remaining sheep. Questions added the additional information that the pastures still looked well and that the soil of the pasture was perhaps 51mm deep and was very dark where it originally it had been very light colored.

It appeared to me that the yearly topdressing of superphosphate, by offsetting a phosphate deficiency had, with the clovers and general stocking management, provided a basis for a rapid but unbalanced soil change. It had enabled the pasture to use the mineral nutrient elements of the soil but only from the top 25mm or two. Eventually one or more of the necessary nutrient elements was completely used up. Then since the scarce element restricts the plant use of other elements, the shortage of the missing factor had produced an impoverished pasture which caused the serious malnutrition from which the sheep died. But since the pasture had been able to support the breeding of increasing numbers of sheep for so long, the now missing element must have been previously contained at least in the top 25mm or so of the soil. Therefore it must be considered unlikely that the now extracted mineral existed only in the thin surface layer and so was still in some sort of supply immediately below the present top-soil. Moreover the type of pasture management, with the yearly dressing of superphosphate which was spread on the ground without any disturbance of the soil, would also have inevitably resulted in faulty soil aeration, which would worsen any other bad effect.

The scientists were at that time conducting tests to try to determine just what the missing minerals were, but the tests had not yet provided the answer, and the solution of the problem was immediate and urgent. Of course, once the missing elements were replaced then the corrective effects could be relatively rapid.

My experience suggested that the improved pasture area should be immediately cultivated with a chisel plough, preferably on the Keyline pattern, and not more than about 75mm deep. The weather was still warm at the time and the moisture conditions in the soil were satisfactory and could improve.

The expected results from the drastic aeration cultivation, which would extend no more than 25mm into the subsoil, would be (1) the rapid development of the soil-life toward a climax which would be well sustained in the warm moist and aerated soil by the abundant organic material present near the surface; (2) those happenings would start reaction in the soil and in the shallow subsoil and result in changes to and reorganisation by humus formation of the minerals presumed to be in the subsoil including those now critically scarce in the topsoil and (3) the first regrowth of pasture would send roots down into the now aerated and moist subsoil and that effect would be followed by the transformation of the subsoil into a soil. The pasture could be affected to the extent of at least partially remedying the deficiency in the first regrowth after the treatment. Soil reactions in such condition could be very rapid indeed.

No one of the group spoke but all looked towards the grazier. His comment was, "That's what we will do!" There has been no other trouble on that particular property, although I did not see the grazier again for 10 years.

EFFICIENCY OF MINERALS ADDED TO THE SOIL

What has been said in those anecdotes is that minerals deficient in a soil may be supplied by one of two methods: (1) From the soil itself by way of soil improvement including deepening of the living soil as has been described; and (2) from the bag of bought minerals. But the fully effective use in the soil of the minerals added will depend on the dynamics of the living soil and its continuous production of its own humus. So the methods of mineral use should include a program of "self improvement" of the soil, so the minerals are then processed by the soil for sustained fertility.

The processes in the soil by way of which the minerals are broken down for plant nutrition, and the details of how soil minerals are taken in and become part of plants and animals do not always conform to the laws of chemistry as demonstrated in the laboratory. Nor do explanations based on those experiments represent the facts of the soil-plant nutrition processes. For example, the early and still too general belief is that the water solubility of fertilisers is the measure of their feeding value to plant. Yet some water soluble fertilisers are quickly washed out of the soil, while others become insoluble in the soil and so useless. It is still the general conception that the water movement from the soil into the plant roots, through the plant stems, branches and leaves and its transpiration to the atmosphere, is the stream which carries the nutrients for the plant.

Some American experience

Dr William A. Albrecht, Emeritus Professor of Soils at the University of

Missouri, whom I visited at his university in 1958, told me of experiments and tests which showed that plant nutrition could continue normally when there was no transpiration of water from the plant, and the "stream" had stopped. Further, that plant nutrients could move into the plant from the soil or from the plant back to the soil, independently of the water movements in the plant. Here are his own words on those two matters:

"As the first fact, plants will grow and their nutrients will move normally from the soil into the roots without the evaporation of water from the leaves. A potted plant, enclosed in a water saturated atmosphere with carbon dioxide under a glass bell-jar in the light, will grow normally. That fact tells us that while the transpiration stream is halted because the saturated atmosphere will not take any water or evaporation, the fertility elements are, nevertheless, flowing into the plant from the soil. In research at the Missouri station, some soybean plants were grown on soils of such low saturation of the clay by calcium, that the total of nitrogen, phosphorus and potassium in the complete crop of tops and roots were less than those of the planted seed. Such facts tell us that the fertility elements may flow out of the roots, or in the reverse direction of the flow of the transpiration stream of moisture."

It was interesting to learn from Dr Albrecht that the Missouri station where I spent four days with him was the second oldest agricultural research station in the world, only Rothamsted in England being older.

The aspect of those discussions on soil and plant nutrition, which may be of considerable value in learning how to manage soil, is whether a sufficient appreciation can be gained of the soil processes of plant nutrition to cause us to reorient our thinking about soil management.

CHAPTER SIXTEEN

Construction of a farm dam

THE PURPOSE OF PROVIDING a discussion on the design of dams for relatively small area water storage is that there is a serious need of it for farmer and grazier and for others who work with them on the land. It is important that a man who is anxious to do the best with what he has in the way of land and its water association be able to think practically on farm dams, and not have a seemingly insurmountable blind spot which may prevent him thinking constructively about improving his land.

A farmer can very clearly see that a channel for water control which is suitable for a certain piece of country is a practical and simple sort of thing which does not in any manner overawe him with its likely cost or with its construction.

But it is a different matter when it comes to interpreting and visualising the same thing on his own land, or deciding on its size. He can see for the first time an irrigation channel working in undulating country and irrigating quickly and economically and then may immediately determine that it would not work for him because his land looked very different, when indeed his particular conditions may have been even more favorable for exactly the same procedure. I have seen and heard those reactions on many occasions. And then there is the reverse effect; a farmer whose land is very flat may see a large flow of water irrigating on undulating country and decide he must have it. He remembers that the channel fall is 1 in 300 and he may select a spot for the start of an irrigation channel on his own land near a dam or a natural waterhole from which he could pump the water. Of course, he finds when he starts taking sights with a level that there is not any 1 in 300 fall on his land at all — it is much too flat. If he persisted in attempts to transpose a technique completely out of its land-shape setting, by excavating the channel as he had seen it, but now with the necessary very flat fall, he would soon find the water would behave very differently. It would not be possible to get the water out of the channel, but merely cause it to spill from the channel and from both sides of it, all over the place, and likely to create a swamp, but certainly not a valuable paddock of irrigation land.

All those matters relating to the various land forms and their associated modes of water control and handling may be, from what is already said, much easier to envisage. With the aid of the channel illustrations of this book and with the special applications described for them such essentially simple matters for those with even a little direct experience of them could be resolved for the interested owner. But there remains to be clarified the dams for water storages which will apply with such much wider views of farm water resource use.

Farm dams for irrigation

The first purpose of the newer enlarged farm dams which these pages envisage remains basically the same as for present ones. They are for holding water when it is available but cannot be used, against the time when it is not available, but could be used. Heretofore farm dams are for stock watering, although many more have been built for the dual purpose of irrigation and stock water since irrigating from dams on Yobarnie began 30 years ago. But as the relative requirements of water for growing all the feed a beast may eat to the quantity it will drink is about 300 to 1, the size ratio for the added purpose of irrigation is likewise very great. It will range from 40 or 50 to 1 up to as high as is practically and economically feasible.

Again there are these other comparisons: (1) It is rarely envisaged in the provision of stock water facilities that all the water available will be required or needed, as stock water is not the limiting factor in production, whereas irrigation water is for the express purpose of increasing production, and therefore the volume of it which *is* available is the limiting factor in production. (2) Stock water dams only rarely require all the available water from rainfall runoff, stream flow, and/or from underground, so siting them is by no means a critical matter in those water resources. But with dams for irrigation, the most efficient siting in their relationship to both the land shape and to each other, is of such significance for controlling and for using the water to increase profits, that the manner of those things may dominate all development plannings, as has been seen already. (3) In cost the stock dam is not of so much moment and usually ranges from below $500 up to $2000-3000 although there are quite a few of them constructed in dry and hot conditions, which are much more costly. The controls along for releasing the water from an irrigation dam will often equal and exceed the higher of those amounts, and the earthworks cost additionally from $2100 to $9000, and on rarer occasions even higher.

There is no widely accepted and used classification for the various type of water storage structures which will be employed for holding farm water where it belongs, principally because, in relationship to their value and to their importance, the subject of them has been almost completely neglected. Professional engineers, with their sights set for only the bigger projects, have overlooked them, and naturally so. Government engineers whose job it could be to concern themselves with research and experiments which would promote the maximum practical use of farm waters have never been given the directive or the money by any government to do either. They are usually not permitted to depart from the orthodox, or even risk one failure which could at least provide some new and valuable knowledge.

There is still in this country the illogical division of authority in agricultural matters which permits the soil conservation officer on the one hand to build a dam or construct a channel on a farm only if it is a legitimate soil conserving measure, and on the other hand restricts a water supply engineer so that he may build a dam only for water storage, but not if it interferes in soil conserving. How could even the best planned farm development stand such nonsense?

A farm dam is a water holding structure consisting of an earth wall which is erected across a valley, and which backs up the water flowing to it, into the valley above the wall. A smaller portion of its waterholding capacity may be held in a hole from which the earth of the wall, or part of it, was obtained, while its principal water storage is held up by the wall over the valley floor and extending upstream according to the wall height and to the slope of the valley. Therefore the main water capacity of the dam is held above the land immediately below it and then, with the aid of a pipe through the wall base in the valley bottom, it can be released on to the land below.

The sites characteristic as between one and another, and with a wall of constant height are favored (1) by a flatter valley floor slope, which determines the distance the water will be backed up in the valley, (2) by the site for the wall being shorter than other sites in the same valley and (3) by the valley floor shapes being wider and flatter in cross section of the water side of the wall as against a narrow "pinched-in" valley shape.

Many stock water storages are simply excavated holes where the water is held below the level of all the immediately surrounding land. Those, usually termed "tanks", are rarely made large enough to carry a sufficiency of supply for irrigation. Such a tank, if dug for the storage of irrigation water, would be very costly as the ratio of water storage capacity to the earth moved is about 1 to 1, whereas that ratio for a dam will usually range from 5 or 5 to 1, and on many occasions better than 10 to 1. Of course on the flatter lands the water-earth ratio may be outstandingly favorable as for instance where a relatively inexpensive wall may back up water 150m or 300m (500 or 1000 ft) for each 0.3m (foot) of height of the wall.

Farm dams which are to store all the available water of a farm however, permit of certain classification relating to their uses and to the shape of the land on which they are constructed.

Other types of farm dams will be discussed elsewhere.

Types and features of a farm dam

The two principal parts of a farm dam are (1) an "earth wall" which holds the water back over (2) the "pond area". The features of the earth wall are its "crest" or roadway on the top of it, which is wide enough for easy maintenance and for equipment and vehicles to pass over it; its inside (water side) and outside "batters" or sloping sides where the degree of the slopes is described as for instance 1 to 2, which means a fall of 0.3m (1 ft) vertically for 0.6m (2ft) horizontally; its "spillway" overflow or by-wash which carries away from the dam the excess water flowing to it after it is filled; its "freeboard", which is the minimum height of the wall above the bottom level of the spillway, and which may be regarded as the safety margin of the dam when filled. The "foundation area" of the wall is the contact of the wall with the solid earth of the land below it. A pipeline, laid during the construction of the wall releases the water from the dam through the base of the wall in the valley.

The "pond area" of the dam is enclosed by the line of the top water level which coincides with a true contour having its starting point at the level of the spillway.

Site selection

The great importance of selecting the site for each farm dam will be apparent to the reader from what has been said before. But it must be mentioned again here. The site for any dam should be chosen on the basis that the dam can become a part of a total scheme of full farm water resource development. The site should not be thought of as something apart, but as a part of something of much greater import than the construction of a single dam. It must not be allowed to restrict, because of bad siting, the profitable possibilities of the whole setup.

Therefore, the first basis of site selection must rest on the examination and understanding of the complete land-water section in which it lies, and as has already been described.

Even the availability of an outstanding site for a dam, if other dams may be added later, should not change the obvious form which the control of the water resources should take.

The outstanding site could usually be employed additionally for these reasons and in one of the following ways: Since the highest place for any dam in a primary valley is at the Keyline the outstanding site must be below that point otherwise it would be used automatically on its own or as a part of the highest interconnected chain of dams. If it lies below the Keyline of a primary valley but too high to be included in a contemplated lower series of dams it can be used by constructing the dam independently. In that case, for filling the dam, a watergate set in the higher diversion channel could remain open at appropriate times. And if there is no such channel an independent channel for the special dam could be constructed and rise from the top waterline of the dam in the valley and into the rising country.

There is invariably more land above to shed water to a channel which rises into the rise of the country, than there would be if a channel were constructed to rise into the fall of the country.

Constructional features for a farm dam

In previous pages it was stated that a point had been reached where, during the course of pegging the line for a diversion channel, a longer peg had been placed on each side of a primary valley. The line of the two pegs was said to represent both the centre of the earth wall of a dam and the height of the top water level of it. Before a dam can be constructed, both the design for the dam and the proposed site for the dam must have constructional practicability. The design is suitable if it can be built with the equipment available and is suited to the material on hand for the construction of the wall.

The lowest costing dam is only possible if the foundation area is satisfactory and if all the material which is to go into the wall is available either within the pond area of the planned dam, or from the land above either one or both ends of the wall site.

The pegging or marking out of the site (on which it is assumed that the waterline contour is already pegged and the centre line of the wall is fixed by the two longer pegs referred to), proceeds by first placing a few pegs to outline the area of the wall's contact with the land of the site. That foundation is

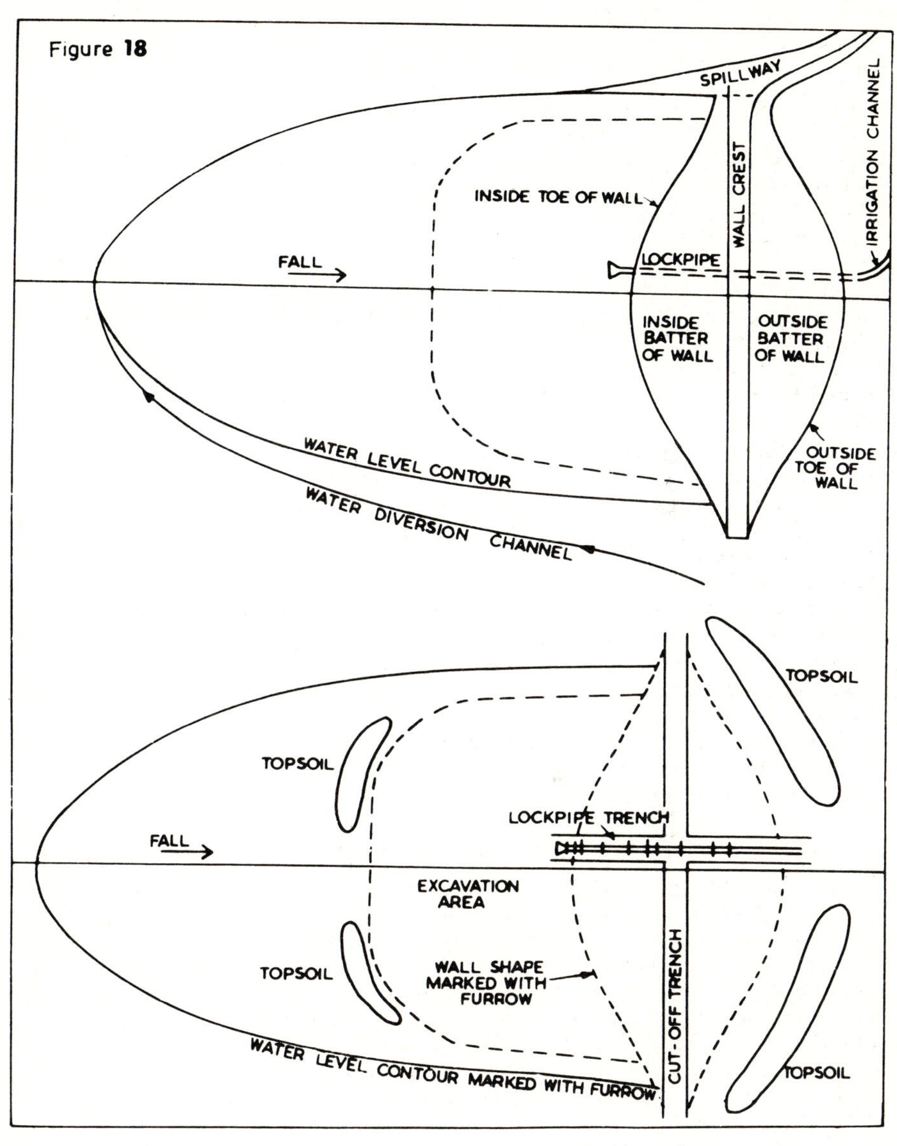

Upper: Plan of a farm dam.
Lower: Plan depicting marking out and site preparation.

examined for suitability by digging holes into it at several points by hand, or with an earth auger or a post-hole digger. The first holes can be 1.2-1.5m (4-5ft) deep, but if the material is satisfactory in the central area of the valley and from one or two holes on the sides of it, other holes need only penetrate through the soil far enough to disclose the same material.

It is always worth finding out something of the local experiences in dam building if it is the first dam for the property. If even the smallest of dams hold water satisfactorily against their walls and above the level of the land, it is quite reasonable to assume that the much larger ones to be built of similar material will also function satisfactorily. Such local information on the suitability of the earth available for dam building can be added to by consulting district water authority officers.

With the determination of the suitability of the wall's foundation area the complete pegging of the site may proceed. In the following pages, reference will be made to the plans in figure 18 depicting in the lower part, the marking out and the site preparation for a farm dam and upper, a plan illustrating the names of the various features. The diagrams of figure 19 illustrate in cross section the possible batters of earth walls.

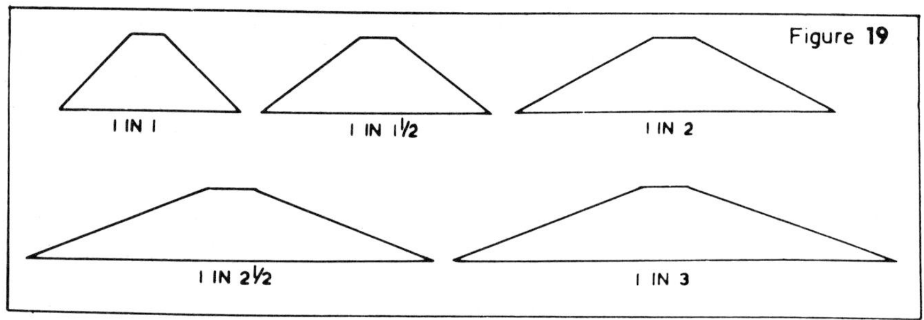

Cross sections showing various batter slopes.

The plan for a dam which is to be constructed will show the detailed measurements of the wall which may be assumed as follows: Total height of finished wall from the centre of the valley below the centreline of the crest of the wall is 7m (23ft) which allows for a 0.9m (3ft) freeboard between the spillway level and the height of the wall. A freeboard of 0.9m is suitable as a general rule except where the width of the spillway is made less than it should be owing to site characteristics, in which case the freeboard would need to be increased.

The width of a spillway will depend on the catchment area of the dam and on the likely intensity of storm rainfall. Since such information may not be available, the following may serve as a guide: Take the square root of a figure which represents four times the maximum catchment area in hectares of the dam and call the answer the width of the spillway in metres. Therefore if the catchment area of a dam is 16ha (25 acres), of which the square root is 4. The spillway width is thus 4m (10ft).

The lock-pipe acts as an additional safety factor if opened up when a dam overflows.

The crest width of the wall, apart from being wide enough for practical farm purposes, also needs to be of a width suitable for working the equipment which will build the wall. A minimum width of 3m (10ft) suffices but if the crest is to be used frequently as a roadway, a more suitable width would be from 4.3 to 5m (14 to 16ft), although the increased width adds to the cost of earth-moving. The batter slopes and the height of the wall and width of crest determine the width of the base of the wall.

Batter slopes

Different slopes for the inside and outside batters are frequently recommended as that is the way with the "big" dam, but in 25 years of experience in the design and construction of farm dams, I have found there is little for, and much against that practice. The one slope for both batters is much more practical for farm dams, which after all are very modest structures when compared with the "book" dams which are for government and large community service. If the farm darm were to be built strictly as the scaled down model imitation of the "big" earth dam, which is 1000 and more times larger in the wall, and as is much too frequently a standard recommendation, its cost would be unnecessarily high.

Many farm dams have been constructed, and served their full purpose, with an inside batter as steep as a bulldozer could climb, and the outside batter as steep as the angle or rest of the earth which was merely spilled over from the wall as it was built. Although most of such dams hold up, that type of structure is not favored, and not only because the rate of failure may be higher, but for the reasons that the crest cannot be satisfactorily formed and would be unsuited to the passage of farm equipment; followup service or reconstruction would be difficult and more costly, and the wall would be unsightly as the outside batter is too steep to grow grass successfully.

I have designed many good dams with 1 in 2 batters, but batters of 1 in 2½ are a little easier for maintenance and for grassing over. In this instance 1 in 2½ batters are assumed, so the toe of the batters would protrude from each edge of the crest two and a half times the height of the wall, which is 17.5m (58ft 9ins). Therefore with a crest width assumed of 3m (10ft), the maximum width of the base of the wall in the centre of the valley would be 38m (127ft 6in), made up of twice 17.5m plus 3m width of crest. The pegging of the wall site could now be resumed.

The first pegs to be placed are for control. A peg is placed in line, on each side of the valley, with the two longer pegs which mark the centre of the wall and crest. The pegs are placed out of the way of the later workings to preserve the centre line and they should not be disturbed throughout the construction of the wall. During the earthmoving, a wall will tend to "wander" a little both up and down the valley, resulting in the moving of some of the earth twice and increasing the cost if control pegs are not maintained, or if they are not constantly referred to.

On either side of the centreline control pegs, others are set to mark the line of each side of the crest. The last pegs are thus also well outside the working area and are preserved in position throughout the work.

Two permanent level-control pegs are necessary, one to preserve the level at which the spillway will "take-off" from the dam and a second for the level as is determined now for the invert of the pipe system to be placed in position before the wall is formed over it. The level of that peg should correspond with the level of the valley centre at the point below the centreline of the wall, or with the valley centre at the toe of the outside batter. If the latter is decided on then the height of the wall is also measured from that level. The two levels are not the same, as the bottom of the valley which will be below the wall is sloping more or less downstream, so the latter level is the lower one.

In connection with the allowance to be made for shrinkage and settlement of the wall of a farm dam the following may be of interest.

During the years from 1944 when we began building farm dams on Yobarnie the customary recommendation for settlement allowance was 10%. In about 1954 some controversy arose over (1) the much lower allowances which I had recommended, and (2) the efficiency of bulldozers for building farm dams.

In 1960 I invited the University of New South Wales and the Water Research Foundation to observe the construction of a dam on Yobarnie and to make any tests they wished. Mr Trevor R. Fietz, BE, AMIE (Aust.), of the university, undertook the work.

The result is reported in the University of New South Wales Water Research Laboratory Report Number 57, titled Research in Soil and Water Conservation Engineering, Progressive Report Number 2, 1960-1961, pages 17-19 and dated January 1962. The report states that the objects of the study were:

a. To observe Keyline methods of dam construction.
b. To assess the general efficiency of bulldozer-only construction in County of Cumberland (Sydney surrounds) soils.
c. To observe the behavior of the embankment after completion and filling.

It should be mentioned that the methods of construction were those I had always advocated and were merely my interpretations of good standard practices, and therefore were in no sense original Keyline methods.

The wall height of the dam was 6m (20ft) and it was constructed as recommended throughout this chapter, and with two bulldozers. The settlement of the embankment from post-construction consolidation was measured by establishing concrete plugs on the crest of the wall and checking their levels at regular intervals. Thirty-three insitu density checks were made and to the date of the report nine of the samples from those checks had been subjected to laboratory compaction tests.

Mr Fietz' tentative conclusions are taken directly from the report . . .

- "In this case the soil compacted at a soil-moisture content close to opmum.
- "With Keyline construction (which means in effect a particular pattern of earthmoving through the borrow area and onto the wall) the compaction achieved with the D6 (Caterpillar) tractor was close to the maximum obtainable with the Harvard Compaction Test.
- "Seven months after completion the embankment had settled 76mm (3ins) in a wall height of 5.4m (18ft), ie a settlement of 1.4 percent. That figure indicates that thorough bulldozer compaction in County of Cumberland soils (earths) makes it unnecessary to provide the customary 10 percent allowance for settlement."

The dam in question has remained in first class condition through the rains of one major flood and two severe droughts.

Shrinking and settlement of the wall needs to be provided for by adding 2 percent to its maximum height which equals 152mm say (6ins), making the total maximum height 7.15m (23ft 6ins).

The pipe system mentioned is not be be placed along the centre of the valley although it takes its level from it. Instead it is placed in an excavated trench a little off the line of the valley bottom and parallel to it and on the downland side.

These two reference level points are transferred with the aid of the instrument to points at the same level on land sufficiently away from the working area so that they are not disturbed or lost. From the pegs so placed, every level and point location for the dam can be readily determined throughout the course of the work.

The area inside the pond line and near the wall site from which the earth will be taken for wall building is marked out with three more pegs. The three pegs are placed at a distance from the toe line of the inside batter equal to or a little greater than the maximum width of the wall, say 43m (140ft) in this instance.

Site preparation

The first part of the "site preparation" can now begin. It involves the removal, out of the way of the future work, of the top soil which now covers the foundation site of the wall and site of the excavation area.

The soil cover over the wall site is placed about 4.5m (15ft) down from the line of the outside of the site, and the soil over the excavation area is placed the same distance up the valley from the line marking the excavation area. The object is to remove only the soil cover, about 76mm (3ins) generally sufficing, and never much more than 102mm (4ins).

As it is nearly impossible to dig such a thin layer evenly with a bulldozer unaided, because the blade will tend alternately to glide over the grass and "dive" in too deeply, the whole area is first cultivated to the appropriate depth. Any farm cultivating implements available may be used, and if a chisel type implement is used two complete cultivations at right angles to each other is the most satisfactory method. Both above and below, a parth is left and not banked up with soil so that drainage down the valley remains open.

The cutoff and the lock-pipe trenches

Next, the preparation of the wall site involves the construction of two trenches. One is the "cutoff" trench, which is to assist the bonding of the wall and the earth below it, and to prevent water movement through the base of the wall. It is located so as to be exactly below the crest of the completed wall and is made with the construction equipment, and of a width to suit the equipment: 3m (10ft) or a little wider is quite suitable. The trench is made deep enough to penetrate 0.3m (1ft) or a little more into the good material below the site. The second trench of the same width and constructed by the same means is for "laying-in" the lock-pipe. It is located clear of the centreline of the bottom of the valley and to the downland side of it, which is the direction toward which the released water will flow. That trench crosses the cutoff trench at right-angles and is dug at the same time. As already mentioned it takes its level, the bottom of the pipeline, from the valley bottom, and therefore is located just far enough out from the bottom of the valley so the excavation into suitable material will bring the trench floor to the correct level. The trench itself, lengthways, is "dead level".

The depth of the two trenches where they cross is not necessarily the same. More frequently the cutoff trench is the deeper.

Before placing earth for the construction of the wall, the material of the foundation area and of the bottom and sides of the two trenches is given a roughup to further assist bonding. A ripper or chisel plough are suitable implements for that work and the depth of penetration need be only 102-127mm (4-5 ins).

The lock-pipe can now be laid and at the same time the first material for the wall can be placed in position, but must be kept clear of the lock-pipe trench.

The lock-pipe system

This aid for improved farm water control and for the volume release of stored water for flow and flood systems of irrigation was developed and first used on Yobarnie in 1945.

It consists essentially of anti-corrosion treated heavy steel flanged pipes and gaskets, steel baffle plates and U-shaped rubber gaskets, screened inflow cone of special shape for the inlet end of the pipeline when laid, and a valve for the water outlet, plus of course, the necessary bolts, nuts and heavy duty spring washers.

Before the use on Yobarnie of the lock-pipe system few farm dams had been equipped with pipelines through their walls, other than the Queensland turkey nest type of dam with its small pipe to supply the nearby stock water trough. At that time a 102mm pipeline through the wall of a farm dam was considered to be unusually large, whereas the present sizes for lock-pipe systems range from 203mm-610mm (8-24ins) diameter with the 203mm system, generally now considered too small.

Although all large government dams have proper water outlets, the provision of the larger pipes for farm dams was considered too difficult an undertaking for the farmer, although official recommendation was that pipes of any size going through the walls of small farm dams should have cement collars cast around them. That was to assist in preventing water from seeping along the pipeline and so eventually washing around it and loosing all the water of the dam and more often, a large portion of the wall as well.

In the lock-pipe system the cement collar is replaced by the more practical steel baffle plate. If in building the wall heavy equipment does happen to knock the pipe, damage that could spoil the effectiveness of the system is unlikely. Also if settlement of the wall or the foundation area of the wall takes place, as often happens, the steel pipe with its adequate thickness is not broken, as would be a more brittle pipe. Anti-seep features are provided in the baffle plates, and the U section rubber gaskets for between them and the pipes. Also lightweight granular expanded perlite "sand" is placed at various places along the lock-pipe when it is being laid, on the theory that if water does start to move along the outside of the pipeline it will also move the light-weight "sand" with it, and tend to block the spaces which could cause the trouble.

The lock-pipe system is placed in position preferably in a wide shallow trench cut by the bulldozer, and immediately before building the wall of the dam over it.

The baffle plates are in two halves and a number of them are set up by being hammered in the earth bottom of the lock-pipe trench. They are spaced from the centreline of the wall to the toe of the inside batter a distance apart equivalent to about one-third the depth of the dam, while from the centreline of the wall to the toe of the outside batter only enough baffle plate halves are placed to hold the pipe up and in line. The plates act in the first place as a cradle to hold the pipes for bolting up. Care should be taken in placing the individual pipes in the baffle plate halves that the completed line is in its right position and that the

230

disposition of the holes of the pipe flanges are such that the valve will be in its proper upright position when bolted up later.

The cradle of baffle plates hold the pipes a little clear of the bottom of the trench to facilitate the proper tightening up of the lower bolts and nuts. Heavy duty rubber gaskets are placed between the pipe flanges, and the bolting up is safely accomplished when it can be seen clearly that these gaskets have been squeezed out a little. The proper tightening of the these pipe junctions is critical and warrants thorough checking.

During the laying of the individual pipes in the cradle formed of the baffle plate halves, the U section rubber lengths are fitted over the pipe contact edge of the plates so that when the top halves of the baffle plates are bolted in position they form a watertight joining of the plates and pipes. (See *Plate 35*.)

With the pipe system laid in that manner and all baffle plates complete and bolted up, filling the trench proceeds.

The more critical phase of placing the earth around and over the lock-pipe is in hand-ramming earth into a tight contact between the bottom of the pipeline and the earth below it. It can be appreciated that no amount of consolidating from above after the line is covered will have any effect on the earth below the pipe.

In filling the trench earth is pushed into it with the bulldozer, from end to end and on both sides of the pipeline, but with care being taken to see the pipe is not covered. The bulldozing is all that is required for the moment, as that earth is now spread by hand shovel over the bottom of the trench and under the pipeline, where it is rammed in firmly. If the earth is spread under the pipe and evenly over the bottom of the trench near the pipeline and a little higher than the bottom of the pipeline, one hit with the rammer covering every place along the pipeline and with the stroke sloping under, will be sufficient for the first ramming. More earth is then placed to cover the bottom of the trench to half-way up the pipeline and another row of rammings, slanted toward the bottom of the pipeline, completes the critical work. If the reason for ramming earth under the pipeline is kept in mind it will be a simple matter to do it properly.

The use of a lightweight expanded perlite "sand" has been mentioned as an anti-seep precaution. It may be placed now by digging narrow holes at the pipe junction and filling them in again with earth and expanded perlite in the proportions of 3 to 1.

The looser earth covering the bottom of the trench should be "worked-down", with a hand shovel, and rammed. But one good ramming stroke for each 152mm (6ins) is all that is required.

The trench can now be filled more with the bulldozer to a little above the level of the top of the pipeline, care being taken to see that the top of all the baffle plates are clearly in view. The earth is consolidated on both sides of the pipeline by travelling over it with the bulldozer and keeping the track on the pipeline side of the bulldozer just clear of the baffle plates. Also, with each of the smaller sizes of lock-pipe systems, up to 254mm (10ins) in diameter or 305mm (12ins) for the largest bulldozer, it is possible then to straddle the system by travelling over it with one track on each side. Care is needed to ensure that there is a sufficient height of earth along both sides to hold the bottom of the bulldozer up from contacting the baffle plates.

When the baffle plates are later covered up with earth, siting pegs are placed centrally at each end of the line and the bulldozer driver is signalled to keep his course of travel on the correct path to miss the now shallowly covered baffle plates. The bulldozer next pushes earth to the system from the sides and at right-angles to the line in such a way that it straddles in turn each individual baffle plate. It spills its load across the trench and pipe and travels over it to a point where the front of the tracks are over the pipe. The driver should be signalled in that work also. The whole length of the lock-pipe and trench is now covered with about 0.8m (2ft 6ins) of earth, and the construction of the wall proceeds normally.

To ensure each end of the lock-pipe system is not covered and lost, the ends are marked with a peg or steel post, and in addition a large old drum is stood up at each end of the open pipe as an additional marker.

The total length of the pipes of the lock-pipe system should be a little longer than the base width of the wall. Since standard pipe lengths vary for different diameters, the length chosen is that which suits the individual lengths and is not less, but could be longer, than is specified.

Constructing the wall

The earth for the wall foundation area is first placed to cover the whole outline of the base of the wall and then levelled off. Good material is used for filling the cutoff trench and the bulldozer should travel it lengthways to smooth out the earth and ensure that each 0.3m depth of earth placed is settled and compacted. Usually the earth dug for the wall will be moist, but if it is obviously very dry from drought conditions it should be watered.

Earth moisture tests were made during dam building on my property where the material was a clay with shale below from which it had formed. Good working condition for evenly spreading the material and for compacting it in the course of building was achieved with a moisture content of 18 and 19 percent.

We have found that throughout the construction the outside or downstream batter of the wall should be continually maintained at its finished batter slope by trimming it to shape twice each day. The batter and all the earth so far placed should be trimmed before work for the day ceases. Also both ends of the lock-pipe are left open to allow possible rainfall runoff to flow out of the incomplete dam.

Maintenance of wall shape

Throughout the whole of the construction of the wall of the dam, the downstream or outside batter of the wall should be maintained at its finished batter line. The inside batter of the wall during construction is not treated in this manner. It starts off on a very flat slope, gradually increasing in steepness until it finally reaches the correct batter on the completion of the wall.

Once the cutoff trench is filled, material is taken from the excavation area and spread across the wall and travelling towards the back line of the wall which was already marked with earth placed during the completion of the excavation of the two trenches.

Supervision is necessary to see that the bulldozer driver does not dig earth from below the site of the wall on the inside of the dam. That is of particular importance in overcoming one of the general faults in farm dam construction. Remember that an irrigation dam will sometimes be filled with water and sometimes empty. The period of greatest stability for the inside of the wall is during the time when the dam is completely filled. The water helps to hold the inside of

232

the wall stable. Its period of greatest instability occurs when the dam is empty. Inside slumping and slipping of the earth of the wall towards the bottom of the dam is sometimes the manifestation of instability. If earth is removed from below the inside toe of the wall during the early stage of wall construction, then fill material will later have to replace it. The result is that a greater length of material that will settle and shrink occurs at the most vulnerable inside point of the wall. If, however, the shape of the land below the wall is preserved in its original form, less only the stripping of topsoil, then there will be a very much shorter length and smaller total area of shrinkage surface and the wall is improved at what is sometimes a point of weakness.

That feature of design is of relatively greater importance in all valley dams as the valley floor slope is steeper. As well as reducing the amount of earth moved, it simplifies both design and construction. The effect is lessened as the valley floor slope is flatter, but it is still of significant importance in design. To be fully effective, good design features should be preserved in the construction of the dam by equally good supervision.

Methods of bulldozer working
Many drivers can drive a bulldozer well by performing accurately all the tasks of cutting and filling required, and though working hard all the time, still move earth at double what it should cost. I have frequently made tests and conducted trials on such matters and always, good supervision and the techniques which follow reduce earthmoving costs by an appreciable amount. In the construction of farm dams the concern is in placing the right material in position as quickly and cheaply as possible, and every technique that aids that end is worthy of consideration.

First, the speed of the engine of a bulldozer is governed to a maximum speed. It cannot overwork or strain itself, so the bulldozer should be worked with the throttle fully open always. Then windrows in bulldozer work are the parallel banks of loose earth left on each side of the bulldozer blade as it moves the earth forward. The action of pushing a bulldozer load forward results in continuous spill of material from each side of the blade. With a large load in front of a bulldozer the movement of the load forward over a distance of, say, 24m (80ft), without digging as the bulldozer progresses, will result in a much smaller load at the end of the distance. Earth will have been lost in spill which forms windrows. But the windrows can be used as an aid to shifting earth faster and cheaper.

The bulldozer should start its run by "grabbing" a big load as quickly as possible, and in low gear if necessary, pushing the load in a straight line and at right angles toward the wall. As soon as the load starts to reduce by spilling in the formation of windrows, then the bulldozer stops and backs up and grabs another full load. That load is pushed forward towards the wall, forming a larger and longer windrow until such time as the load is reduced below the full load capacity of the blade, when the bulldozer rips back and grabs another full load. Modern bulldozers can be equipped with "back rippers", which tear up the earth as the bulldozer moves backward. The bulldozer proceeds in that manner until windrows sufficiently high are formed, which will enable a full load to be transported forward to its final position. The bulldozer is now confined between the windrows, pushing up from six to 12 full-capacity blade loads into the wall site.

Grabbing the load simply means loading the bulldozer blade to maximum capacity as quickly as possible and using low gear if necessary, so that the rest of the trip to transport the full load into the new wall can be travelled in a faster gear. If the full load is not obtained in the first pass over, say, a 6m run, the driver rips back and grabs a load again until the full blade load is obtained. After six to 12 full passes have been made in that pair of windrows the tractor is moved to form a new path and new windrows, with one windrow of the new path partially formed by the windrow of one side of the first pass.

The new windrow site is worked as before. Windrows may be up to 0.9m high.

The windrows should be formed at the centre of the excavating site and moving out first on one side and then on the other, so that when all windrows have been formed and used there is a series of parallel equidistant lines of windrows lying at right-angles to the wall of the dam.

The next task involves the destruction of all the windrows of the first series of passes. The bulldozer starts the second series by travelling with one of the old windrows in the centre of the blade, pushing a maximum load of the old windrow material forward at right angles to the wall. That pass will form new windrows very rapidly, and the bulldozer continues working in the newly formed pair of windrows for the suitable number of passes. The next movement of the bulldozer pushes the next windrow of the first series out on either side of the new pass and continues until all the old windrows are bulldozed out, new windrows have been formed and the requisite number of full loads taken between them. This method is followed throughout the whole of the construction of the dam.

Systematic working and following those procedures will often shift earth for less than half the cost of another method which, though it may appear quite satisfactory and economical to both farmer and bulldozer driver, is much more costly.

The bulldozer, except for trimming the earth and in the final finishing-off of a job, should not travel to the wall site with less than its full load. A bulldozer travelling on to the wall site with half a blade load of earth is shifting earth expensively.

Supervision

Supervision in the construction of a dam is not provided merely by a man watching the bulldozer work. A supervisor, whether a farmer or anyone else, must know first what he wants, and as many people may not have seen a good farm irrigation dam then he should have a plan. The plan should be first studied so the farmer may be convinced of the logic and necessity of every detail of the design, of the methods of work, of the construction details, and of the final finish and the use of the dam. He then should see the driver follows his instructions. A farmer may be somewhat reluctant to instruct a bulldozer driver of wide experience, and think that he should not be told or should not be stopped when he is not performing the work according to plan. However, a bulldozer driver in almost all other circumstances outside farm work works to a plan and under a supervisor, because such methods have been found to produce lower costs and efficient work. Moreover, it is quite unfair to a bulldozer driver to expect him to design and construct a good dam from the seat of his tractor as he goes along. Also, the driver is only on the farm for a short while, but the farmer has to live with the work, whether it is good or bad, for very many years. He should, therefore, get the dam he wants, and he will get it only if the manner of his supervision produces effectively his design of the dam.

A bulldozer is not a completely efficient compacting machine. Nevertheless, the travelling of the bulldozer over the earth as placed does give a measure of consolidation and stability to the loose earth, and can produce a very desirable uniformity of texture in the material. Uniformity of texture is important, as it assures that shrinkage, if it does take place, is also uniform, while cross cracking and longitudinal cracking of the wall will be lessened.

During the construction of a dam, continual waves of earth in the wall site are to be avoided by the bulldozer travelling forward over the earth as it progressively drops its load in the wall site, and as the driver slowly raises the blade. The fill material must be continually smoothed off and shaped up.

The construction of a farm irrigation dam may occupy as little as two days, or may take several weeks, depending on its size, the implements used, and digging conditions and weather.

In dam sites with valley floor slopes which permit all the excavation material to be taken from above the level of the lock-pipe, care should be taken to see that the bulldozers do not big below that level. Those spots above the level of the lock-pipe which would trap rain water are also to be avoided. The full depth is to be maintained only in the lock-pipe area, so that rain water will drain to and through the lock-pipe. Before finishing for the day, the windrows left in the excavation area from the day's work should be pushed into the wall area and trimmed. The whole area, including the wall, then will carry the minimum of loose earth and provide good drainage from the work to the lock-pipe.

Preparation made at the end of each day's work should provide against the possibility of damage by heavy rain falling during the night. Some centimetres of rain could fall on the construction then and not prevent work on the following day. However, loose earth on the walls and on the excavation site would, in the same conditions, absorb much water, create ponds, and possibly hold up the work for days. The lock-pipe should also be maintained in an open, free and working condition at all stages of the construction, and should be checked and cleared of earth each day before work ceases.

If rain has fallen on the work area the whole of the wall section should be travelled and ripped up if necessary before starting again, so that continual bonding of the earth as it is placed in the wall occurs satisfactorily. Windrows are then formed again as usual, when the deeper, somewhat drier earth will become well mixed with the smaller amount of wet material, so uniformity of moisture content and texture is stil maintained. In many types of earthwork moisture conditions are maintained within precise limits, but generally the only occasions which may be troublesome in the construction of a farm dam are when the earth is very dry and does not bond propely, or when it is in an over-wet condition that causes clays to ballup and leave air pockets in the wall.

Progress of work

During the construction of the dam all bulldozer paths should be at right angles to the wall and be particularly so in the early stages of the work. After the task of laying, tamping and filling the lock-pipe trench is completed, the formation of the wall should continue with the bank at the highest level over the central low area of the valley bottom. From that stage to the time when finished levels are being considered the central area of the wall should be maintained as the highest point with a slope of about 5 percent along the length of the wall from the centre to each side. When that wall line is maintained during construction the low portion of the wall is always well away from the area where the maximum earth had been deposited. In a valley of uneven section, for instance, where the one bank of the valley at the wall is steeper than the other, the lowest point of the rising wall at any time will be the end of the wall on the flatter side of the valley and at the greatest distance from the main earth fill. Water would flow over the wall here if the partially built dam were flooded. The low spot acts as a safety valve, thus protecting the main fill area of the earth wall.

In a sudden heavy downpour of rain the lock-pipe would also be flowing full of water and as soon as heavy rain has ceased it would quickly reduce the water level to below the overflow height and then empty the dam. Damage would be at the minimum, even though no expensive safety precautions had been taken.

Dams with only small natural catchments would hardly be affected by heavy rain. The lock-pipe provides full safety as floodwater would only rise to a few metres of depth for a short time. That safety factor is aided by the fact that the diversion channel which will later help to fill the dam is not constructed until the wall is completed, so that runoff is restricted to the small natural catchment. It can, however, be even more restricted if need be by constructing a smaller diversion channel around the area of the dam.

A well-planned and supervised job will look correct throughout the job. Some bulldozer drivers like to concentrate on one spot to bring it up to its finished height, but that is bad practice, as it tends to work against uniformity of texture and proper bonding of the earths in the wall, Further, areas of compacted stabilised earth are likely to be placed adjacent to areas of very loose earth. Shrinkage later would form large cracks between the different textured materials. There is a tendency also for bulldozer drivers to push all the earth of the blade-load into the wall and leave it as a loose mound. That can be avoided by the drivers starting to lift the blade at the correct position on the wall so the earth is distributed evenly. Again, there is a likelihood or earth being carried forward too far on to the wall site, the result being that much loose material is spilled over the back of the wall. Loose material in a finished wall tends to absorb a lot more rainwater than the rest of the wall, which, by increasing its weight, could cause sliding and slumping of the rear of the wall. The back batter of the wall should be maintained throughout by trimming with the bulldozer as it becomes necessary.

Sometimes a bulldozer driver "rushes-for-height", which often results in a concave line up the wall. The line up the wall should always be a straight line. A concave line encourages slumping of the high point on the edge of the crest of the wall, and once that has started the extra weight on the material below causes a movement of earth which in the worst circumstances could result in a later partial failure of the wall.

Throughout building the dam the marking pegs should be maintained in position by lifting them out of the bulldozer path when it is necessary, and by the farmer stepping three or four paces out and lining up the position of the peg between himself and another peg, so that after the work the old peg can be put back in its correct position.

Spillway

The spillway of a dam, like all other features of good farm irrigation dam construction, has to be correct for the dam to be a good one. Usually the construction of the spillway will produce a surplus of earth which is then used in the building of that end of the wall.

The batter between the end of the wall of the dam and the bottom level of the spillway should be about 1 in 3 or flatter, so that when grassing the dam site, wall and excavation area, convenient travel with cultivating equipment is possible. A section of the spillway therefore will show a batter slope of 1 in 3 falling from the end of the wall of the dam to a dead level spillway bottom to the rising land on the high side of the spillway and away from the wall of the dam.

Shrinkage allowance, freeboard height and spillway size therefore provide that, on the completion of the dam and its subsequent settlement there will be 0.9m of wall everywhere above top water level at the point where water starts to flow out of the dam and through the spillway. The design of the spillway is such that the type of flood likely to occur any time in 50 years would be by-passed with the spillway carrying little more than 0.3m of water across its full width and when that happened there would still be another freeboard of 0.6m to compensate as a safety measure for bigger floods.

Larger spillways are necessary for farm dams built in the lower parts of large primary valleys and for creek dams, as they usually have considerably more catchment area than a dam higher in a valley. To secure the necessary width of spillway with a dead flat floor, considerable material may have to be excavated into the rising land near the wall of the dam. In those circumstances an appreciable amount of earth may have to be moved, that is, earth greatly in excess of the needs of the spillway bank. The construction of the wall of such a dam then is designed so that the earth of the spillway is used in the construction of that part of the wall adjacent to the spillway. It is advisable to construct the spillway before the earth in the centre of the wall approaches its finished height and when there is still plenty of wall area unfilled and available for the use of the spillway material. On occasions where the spillway of a dam involves very considerable earthmoving, the construction of the spillway may be completed by placing the excavated material into the wall site immediately the site preparation is completed.

In spillway construction earth has to be excavated down to a specific level and supervision should ensure that the earth is not excavated too deeply, because it would then necessitate the filling of areas of the spillway with earth that would not be in as good condition as the stable undisturbed material.

Final Batters

With the earth for the wall being constantly moved in properly designed wind-rows at right angles to the wall, the site is in a condition to be examined continually to ensure that the cheapest digging earths go into the wall. As the work proceeds, areas of material somewhat harder than that of the general digging conditions may be encountered. Such areas are then studied to determine whether cheaper earth can be obtained by going back another 4.5m or 6m for earth, or whether cheaper earth is obtained by persevering with the cutting and digging the harder materials. Outcropping hard rock may be encountered and it should not exceed 25 percent of the earth in any blade load. Bulldozers with back rippers are capable of making fairly light work of reasonably tough materials.

The cost of moving a particular quantity of earth is related to the distance it has to be moved, so length of movement of earth should be considered as the work proceeds.

With the outside wall batters strictly maintained on a straight line and at the correct slope, the front wall will gradually steepen from very flat slopes towards the final finished batter. Centreline pegs, lined up with the pegs on either side of the valley, should be placed at intervals during the final stages of construction and measurements should be taken and marked with temporary pegs to show the finished width as well as the final height of the wall. A continual check with a level and measurements ensures that the right amount of material is in the right place and excavated from the correct position in the excavation area. The centre are of the wall is first finished off to its final height, crest width and batters. The construction of the finished batters, the heights and top width of the remainder of the wall is maintained towards the sides. Final batters should be even and continuous throughout with straight lines up the wall. A slightly steeper slope in part of the wall always has a tendency to be more unstable than the rest of the wall. Slight movement of earth may take place, the parting on either side of the movement being represented by a steep crack, which tends to allow movement of the flatter battered earths on each side of the steep area and thereby assisting the movement of the steep material and progressively worsening. Such movements tend to reach a point where they stabilise themselves, although not necessarily so.

When the material has all been placed and the wall trimmed to its proper top width, height and batters, the final finishingoff starts. The windrows left in the bottom of the cut may be flattened, but it is not necessary or advisable to move all loose material from the excavation area. However, it is important that the excavation area be smoothed into a natural shape to conform to the valley area of the whole dam. When that is finished, the raw earth is cultivated with a chisel plough and then covered with the soil which was previously moved from the surface of the excavation area. Some of this soil is used also to cover the inside of the wall.

The soil which was stripped from the wall foundation area is brought up over the wall to cover the outside or downstream side of the wall and the top of the wall. About 51mm of soil cover is all that is required.

The crest of the wall should be finished off with slightly rounded edges. If it is finished off haphazardly with a bulldozer there is likely to be a small windrow effect left by the blade. When rain falls that could cause little ponding areas, which eventually break in one particular spot and water flows down the wall in a concentrated stream. More rain falling on the wall takes the same path and sufficient damage could take place with a few millimetres of rain to spoil the appearance of the new wall and necessitate some repair work.

The area of the dam is cultivated with a chisel plough in a single-run cultivation about 76mm deep. The cultivation parallels the water level contour downwards (Keyline cultivation), so that flow water later spreads as it flows into the empty dam. Next, the wall and the whole of the site is sown with the regular pasture seed mixture and combined with a dressing of starter fertiliser.

Hand finishing of the top of the wall to leave a good shape and to aid the germination and growth of the grasses is well worthwhile. Slight ridges of loose earth can be raked out.

Finishing off the Lock-pipe System

The volume strainer is coupled up to the cleaned surface of the flange of the pipe on the inside of the wall with a rubber gasket between and with the straight edge of the volume strainer downwards.

The inclined section of the lock-pipe is made to point upwards in the up-stream direction so that the upward tilt acts as an additional safety to preserve a free and full flow of water in case there is a slight slip of wall material through any cause.

The strainer opening is nine times the lock-pipe size, and screened by heavy mesh. The speed of the water flowing into the volume strainer when the lock-pipe valve is open is therefore very much slower than the speed of the flow through the lock-pipe so that any rubbish which can enter the strainer will flow out through the pipe.

The outlet valve should be coupled in the same manner as the strainer but on the downstream end of the lock-pipe and tightened up and closed. All surfaces, gaskets and flanges should be clean and no earth left in the end of the lock-pipe. As care has already been taken in the laying of the lock-pipe to ensure the holes match up with the flanges of the lock-pipe the valve will fit in an upright position.

Valves may be provided with a 51mm outlet on the water side of the valve closure so that water is always available for such items of smaller supply as stock troughs.

CHAPTER SEVENTEEN

Dams for flat land irrigation

ALL FARM DAMS which are constructed in primary and secondary valleys have design and constructional features similar to that of the dam described in Chapter 18. Such dams will differ widely in their capacities even with similar wall sizes and heights, the big factor in the difference being the important one of the slope of the floor of the valley at the site of each dam. For instance, assuming a constant wall length of 120m (400 ft) and a constant water depth of 6m (20ft) at the wall, a dam with a valley floor slope of 1 in 15 would have a length from the wall to the water line in the valley bottom upstream of 90m (300 ft); with a valley floor slope of 1 in 30 the length of the dam would be 180m (600ft), and with a valley floor slope of 1 in 100 the length would be 610m (2000 ft).

Minimum shape for a farm dam

A dam may be considered to have a minimum shape if the length of the dam is equal to the length of the wall. In the above example a wall length of 120m in a valley of 1 in 15 slope makes the length of dam only 90m, so such a dam is not of minimum shape and would not be considered. The other two examples would be considered of satisfactory shape, but not if other sites were even more favorable.

The highest possible sites for valley dams are those having their top waterline coinciding with the Keylines of the primary valleys. At such places are found the widest ranges of choice, where many positions at the Keylines are completely unsuitable for dams while another valley's Keyline site may be very good. Lower down in the same primary valleys the suitability of possible sites for dams does not vary nearly so much. The particular site chosen, where a chain or series of dams in the middle or lower positions is to be used, will be most often suitable indicated by the diversion channel and by the depth selected for the dam.

The depths of farm dams

The depths of farm dams and the height of the walls are determined individually for each site. The principal considerations are (1) the slope of the valley floor and (2) the water capacity of the site. Provided there is ample water available from the natural catchment or which can be brought in by diverting it from outside, the dam should be as large as possible and be near the maximum economical wall height. Walls over 9m high should be individually designed on the more orthodox engineering lines.

A good safe maximum for a farm dam in medium undulating country is a 6m (20 ft) dam with a 6.9m (23ft) wall. Of course many sites may warrant higher and larger walls but even so it is advisable to first check the site by pegging for that size dam. If a dam 6m deep is considered to be a good working maximum then 3m (10ft) deep is similarly a good working minimum and the large majority of farm dams for irrigation would fall in that range.

From what has been said in reference to the valley floor slope of the site for a proposed dam it will be appreciated that, as the land considered becomes flatter, the water storage requirements tend to be satisfied with lesser wall heights. Also the flatter valleys tend to become wider and so the walls for dams in them will be longer. There a 6.9m wall becomes an increasingly larger undertaking. In such cases the first checking of a proposed flatter site would be made by pegging it as for a dam 3m deep or a wall height of 4m (13ft), and by increasing the size if that proves smaller than the requirements.

On land which is generally considered to be flat and dry, the primary valley shapes of Keyline still persist but may not be obvious to the eye. But on such land the flat primary valleys in flat surroundings can often be used for the temporary storage of heavy rainfall runoff. The walls for such storages need be little more than long low barrages up to only 1.8m (6ft) high. The water would be used as soon as profitable on a paddock fenced and specially prepared for it, and be released from the storage quickly by a large lock-pipe or, in some circumstances, by watergates. Such a water storage may provide water for irrigating only three times before the water is all gone, but it could still be a most profitable undertaking. In the bottom of the dam a deeper and smaller excavation could remain as a stock watering point.

A farm dam in a creek

The various valley dams are generally simple construction jobs but the damming of creeks involves wider ranges of site conditions. The damming of some creeks would be beyond either the capacities or economics of the farmer.

While on many occasions some streams which have very large catchment areas can be controlled and have water diverted from them for farm use, the smaller creeks are more likely to be suitable for such uses.

Since a water licence is necessary for creek dams it is as well, in the first instance, to discuss the site possibilities of a proposed creek dam with the local officers of the government water authority.

The basis of the design and of the construction details for a creek dam involve first the selection of a spillway area which will allow overflow water to leave the dam and return to the creek lower down without damage to the immediately adjacent land. Second provision is made for the control of the water which may be flowing in the creek during the construction of the wall, and third the preparation necessary for the site for the wall is determined. Pegging the site itself is done as with a valley dam.

Selecting the site for the lock-pipe is also more critical than for a valley dam. Preferably it should be located by excavating the lock-pipe trench away to one side of the flowing water and with the level of the bottom of the trench a little lower than the creek. The trench is kept isolated from the flowing water until the lock-pipe is placed and properly covered up, when the flow of the creek is turned into it. With the flow controlled the preparation for the wall site is done as for the valley dam.

The selection of the excavation areas which will provide the earth for the wall of a creek dam is also a little more critical than for a valley dam. If the material can be obtained from near the wall and below the waterline of the finished dam, so much the better, but frequently it must be procured elsewhere. Generally the wall of a good dam site on a creek is associated with the end of a longer primary ridge on one or both sides of the creek. The spillway may be cut out of the end of one of those ridges, and as well the same ridge could supply all the earth for the wall.

In the construction of the wall itself, which follows that for a valley dam, more care should be taken to protect the main bulk of the wall from possible flood damage, so throughout the work the wall is made to slant along its length from the side of the creek with the steeper bank. Flood overflow would then occur farther away from the main section of the rising wall. With the lock-pipe open and free during the construction of the wall, the maximum protection is then assured.

Frequently water will have to be pumped for use from a large creek storage, therefore a site for the pump and its suction line should be prepared as part of the construction work. That should ensure all or most of the stored water can be pumped from the dam and from the one permanent setup of the pump, and not by having to move a pump down to contact the falling water level.

Non-valley dams

In some circumstances where valuable water flows to waste there may be no valley or creek site for a dam. There are two designs which may apply and which will provide dams of economical cost.

First a contour dam may be used to advantage on slopes which contain no valley form. That dam is essentially a long earth wall of medium height constructed from earth which is excavated from immediately above the main wall, and with wing-walls made to taper up the slope to above the water level of the dam.

In the flat lands all design features of the dams are flatter; the dams themselves are shallower, the water diversion and irrigation channels are both flatter, but the irrigation channels are built up so that the water flows above the level of the land. The land to be used for irrigation is also flatter and all the flat land methods of irrigation can be used from the supply held in contour dams.

The critical design feature of this type of dam, other than the all-important one of climate and its associated run-off, is that of slope. Contour dams can be constructed on slopes ranging from 1 in 25 (4 percent slope) to 1 in 100 (1 percent slope). They may be classed or named "straight" "inside", and "outside", according to their general contour shape. A straight contour dam is one whose wall follows a contour curving around a flat ridge shape, the dam being on the inside of the curved shape. An outside contour dam is one associated with a flat valley formation where the water lies on the outside of the curve of the wall.

A contour dam should be located as high on the property as convenient. There must be runoff and sufficient catchment area above the water diversion channel to fill the dam. As to size, it may range from 24,000m^3 (five million gallons or about 20 acre feet) to 120,000m^3 (25 million gallons) or more. To bring the matter to a practical consideration, we may assume a contour dam is to be designed with a capacity of 100,000m^3 (80 acre feet) of water; that the slope of the land is 1 in 50 (2 percent slope) that the depth of water at the inlet to the lock-pipe is to be 3.7m (12ft); that the land shape contains large low forms only; and that the contour shape of the dam is "straight". That capacity would require a wall about 274m (900 ft) long. Water 3.7m deep on a 1 in 50 slope would place the water line up the slope 3.7 multiplied by 50, or 185m (600ft), the dam thus having an area of a little over 5ha (12 acres). The average depth of a contour dam is somewhat over 50 percent of its full depth, or about 2.1m (7ft) in this case, so that the required capacity is satisfied by that general size.

In the medium-size farm dam, a suitable freeboard height is 0.9m but the circumstances of design in a contour dam suggest that the figure be reduced to 0.6m or a little less. There is no part of the wall of a contour dam that represents the main bulk of the earth, as is the case in the valley dam, and a failure of part of the wall is not nearly so serious a matter as in the valley dam. Moreover, the inflow of water to the dam is readily controllable.

Those facts also suggest that the minimum or cheapest construction methods may be used in building the wall, and also that, with the lower wall height, the wall batters may be steeper. Wall height will then be 3.7m depth of water plus 0.6m freeboard, and as minimum construction methods are to be employed no allowance for settlement and shrinkage will be made. The constructed wall height is therefore 4.3m (14ft).

The dimensions of the wall section are as shown on the plan and section in figure 20, page 203. With the constructed height 4.5m, the width of wall shape at the base is 20.2m (66 ft), with batters 1 in 2 and the crest width is 3m. The lock-pipe will be placed into solid ground and there will be about 0.6m of solid earth above the lock-pipe level on the inside of the wall. At distances of 24 and 30m (80 and 100 ft) from the inside toe of the wall there will be 1.2 and 1.5m (4 and 5 ft) of earth respectively above that level and more than sufficient for the wall without digging earth below the inlet level of the lock-pipe.

The water diversion channel for a contour dam, like those for all dams, does not fall directly into the dam, but is constructed right along and above the dam, reaching water level height at or near the spillway end of the dam. The channel may be much flatter than the fall generally employed for the valley dams. A flatter channel has less capacity, so that the channel needs to be of larger section. The channel should fall in the down-land direction.

The position of the lock-pipe may be in any portion of the length according to where the water is to used. If the area of the slope immediately below the dam is to be irrigated, the lock-pipe is placed in the main wall at the end where the diversion channel first reaches the dam. In other circumstances it will be placed in the opposite end.

The price a cubic metre of earth moved in a contour dam of lesser wall height will be considerably less than in the higher-wall dams. The average haul will be less, the push up the batter of the wall is shorter, and more of the work, which is only shallow digging, can be performed in second gear. A reduction of

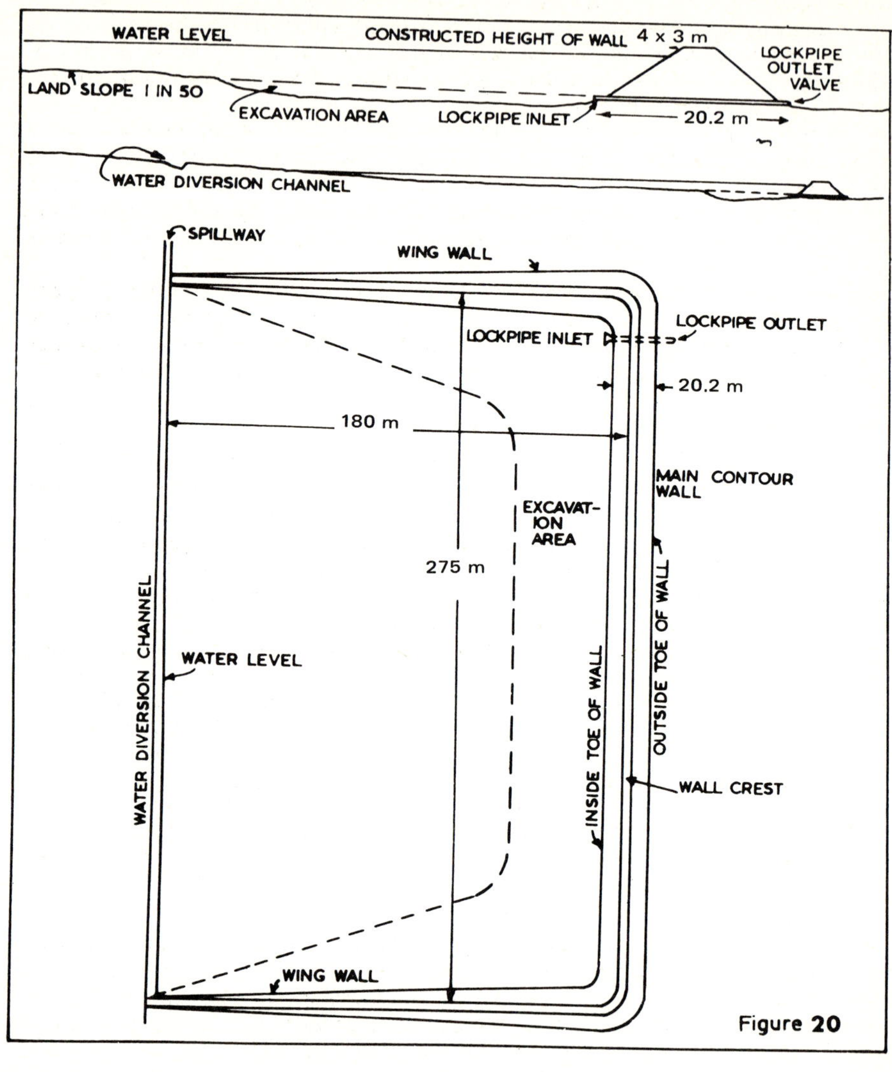

Design of a straight shape contour dam.

Upper: Cross section of excavation area and wall with 1 in 2 batters.

Middle: Section across the dam.

Lower: Plan of contour dam showing wall, water diversion channel excavation area and wing wall.

40 percent in earthmoving costs is to be expected as against a wall 7m high.

Marking-out and site preparation should proceed as for a valley dam, with the clear marking-in of the wall shape on the ground with a furrow line. Top water-line for the dam should also be marked. That part of the water diversion which is along the top of the dam could be first constructed to prevent any run-off into the area of the dam during construction.

A cutoff trench for the full length of the main wall and the two wing walls should be used, but may need to be only a few centimetres deep. Even where it may be considered the cutoff trench is not required, it is still advisable to use a shallow trench, as it helps appreciably in controlling the job and in supervision. The area of the wall site is chiselled along the line of the walls to assist bonding as before.

A contour dam of that style provides a water-earth ratio of about 6 to 1, and the same general structure on land sloping 1 in 100 would improve on the ratio to 8 to 1. Having regard also to the low cost of earthmoving for a wall of such modest height the structure is a very economical one for farm water storage.

A ring dam

Where the slopes are so flat that the contour dam becomes unsuitable, there is no way to store water above ground level other than with a closed wall dam and by pump filling. The most usual source of water supply is a stream, which may flow only after heavy rainfall, and those facts should be noted when locating the dam and designing the dam and its related filling structures. In some circumstances, though a dam must be filled from a certain watercourse, it may be a disadvantage or even an impossible hazard to construct the dam close to the watercourse. In other circumstances and for the sake of efficiency and economy, it may be worthwhile to depart from the ring shape by having part of the wall following a portion of the bank of a watercourse. (See figure 21, page 205.)

There is a general idea that water has to be pumped over the top of the wall of such a dam. However the ring dam may be filled also by pumping through the lock-pipe when suitable.

The construction of the ring or other closed-wall dam follows the general procedures already given. The size of the lock-pipe may be increased according to the capacity of the pump which is to fill the dam, but generally a size of 305mm (12ins) is suitable.

The importance of making proper arrangements for the filling of ring dams, or any closed-wall dam, cannot be overstressed. The full layout should be decided and included in the design of the dam itself. A creek weir diverting flow to a channel and bay, from which the water is to be pumped, should all be constructed and completed as part of the dam construction. The elevation to which water has to be raised in the dam, the pump capacity and power requirement, must be logically determined in relation to the capacity of the dam. Generally, where water has to be pumped into a dam, time is so limited that large capacity, low-head pumps are invariably required. The dam for those reasons should be close to the level of creek flow, so that pumping will take place from only slightly below ground level. The likely length of time available after storm rains for pumping should be calculated against the capacity of the dam, that is to say when the rate an hour of water delivery required has been estimated, a pump capable of that performance against the height of the total lift should be acquired.

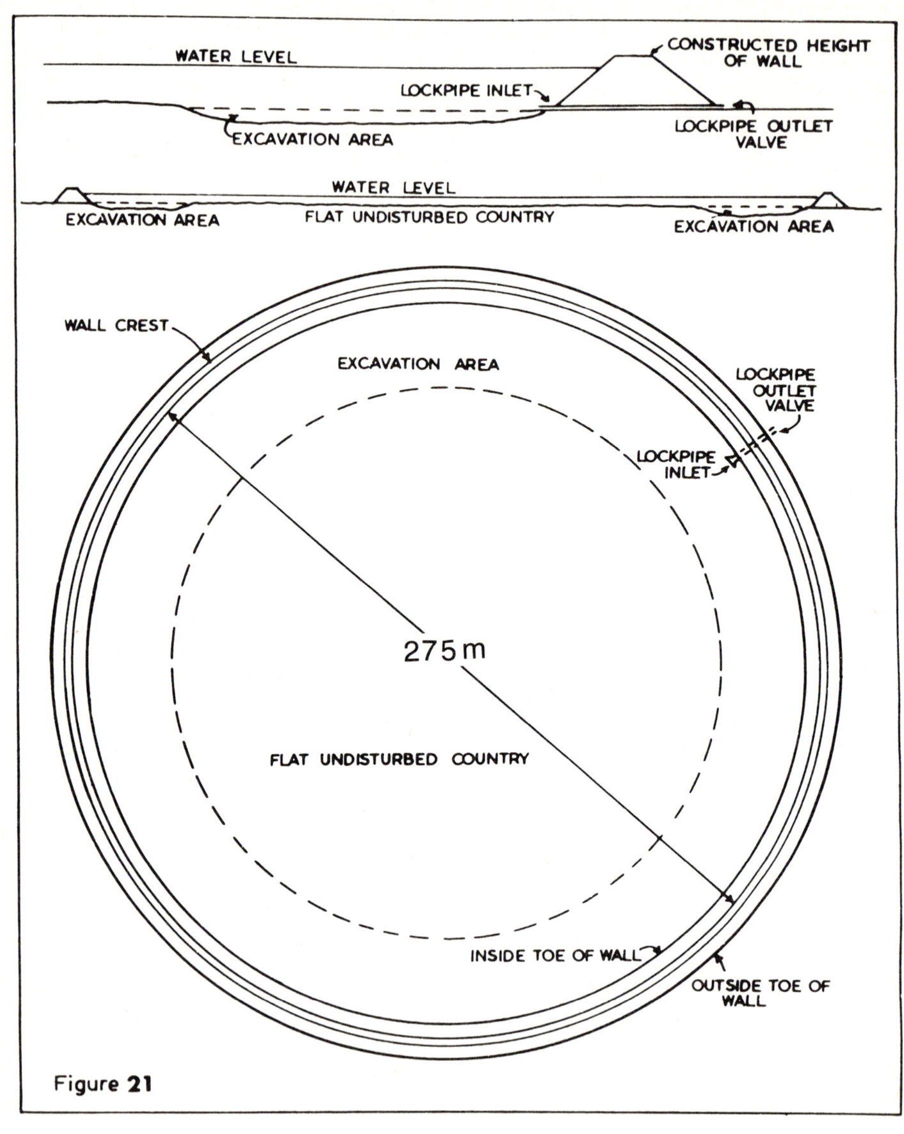

Design of ring dam.
Upper: Cross section of excavation area and wall.
Middle: Section of ring dam.
Lower: Plan of ring dam.

The most suitable dam filling arrangement is a permanently located pump and engine which can be used under the worst weather conditions. A lower initial cost method is to arrange a permanent pump setup so the power of a farm tractor can be quickly coupled to the pump. With the pump in a permanent position and the suction line in place, the setup is always ready for immediate use. A large low-head high-capacity centrifugal pump selected to exactly suit the requirement or an axial flow pump may be used.

The water from a ring dam can be used for many different irrigation systems, varying from flood to any type of spray irrigation.

*   *   *   *

As mentioned earlier, I do not fully subscribe to the belief that food supply will become a critical shortage factor in population trends. Transport and exchange of food supply may fail. The opposite effect, that of over-supply, is more likely to pose a problem of production costs to the Australian farmer. Here the greatly reduced costs of soil improvement for crop yields and pastures that are made by those methods may be of vital importance. However, whether prices tend lower or not, lower costs and higher yields from continually improving soils are satisfactory aims themselves.

The potential of farm water resources is of outstanding importance in Australia both nationally and personally to tens of thousands of individual farmers. Anyone in authority who, presenting broad aspects of the developments of farm water resources for irrigation to the landowner or student, disregards relevant facts and does not mention even the possibility of applying water to land by systems other than "spray irrigation" is doing a disservice to Australian agricultural development.

Irrigation from the water resources of our farm and grazing land should be made as profitable as is possible, and as was mentioned early in this book: "Whether irrigation will produce much profit or result in substantial losses often depends on just the right choice of irrigation procedure."

Index